**Concrete for
Structural Engineers**

Concrete for Structural Engineers

A text to CP 110

C. B. WILBY, BSc, PhD, CEng, FICE, FIStructE

*Professor of Civil and Structural Engineering,
and Chairman of Postgraduate and Undergraduate
Schools of Civil and Structural Engineering,
University of Bradford, UK*

NEWNES-BUTTERWORTHS
LONDON-BOSTON
Sydney - Wellington - Durban - Toronto

The Butterworth Group

United Kingdom	Butterworth & Co (Publishers) Ltd London: 88 Kingsway, WC2B 6AB
Australia	Butterworths Pty Ltd Sydney: 586 Pacific Highway, Chatswood, NSW 2067 Also at Melbourne, Brisbane, Adelaide and Perth
Canada	Butterworth & Co (Canada) Ltd Toronto: 2265 Midland Avenue, Scarborough, Ontario, M1P 4S1
New Zealand	Butterworths of New Zealand Ltd Wellington: 26–28 Waring Taylor Street, 1
South Africa	Butterworth & Co (South Africa) (Pty) Ltd Durban: 152–154 Gale Street
USA	Butterworth (Publishers) Inc Boston: 19 Cummings Park, Woburn, Mass. 01801

First published in 1977 by Newnes-Butterworths

© Butterworth & Co (Publishers) Ltd, 1977

ISBN 0 408 00256 5

Printed in England by Page Bros (Norwich) Ltd

Preface

CP 110 unified three previous codes, made use of research and experience since the previous codes, used some European philosophy, SI units and international nomenclature, and out-dated existing U.K. text books. The writer's objective is to produce a post CP 110 book (including the August, 1974, and February, 1976, amendments) enveloping everything concerned with concrete (properties of concrete materials, mix design, reinforced concrete, reinforced concrete shell roofs and folded plates, and prestressed concrete) required by a student for a British bachelor degree course in civil and/or structural engineering. As the courses vary, and some topics are taken further for more completeness in the book, the book is useful for some of the work of MSc courses in structural engineering and concrete technology, and many of the MSc students, particularly the mature ones, need the modern undergraduate work in the book for revision purposes. For students the book has to give principles and, for completeness and usefulness, applications.

As the applications use CP 110 the book should be useful to the large number of British practising engineers knowing principles and the previous codes, but trying to cope in design offices with CP 110, which is a difficult document to interpret with confidence, and made more so, for experienced engineers, by requiring knowledge of SI units.

Many design tables have been included to help students in their design office classes and to help practising designers with CP 110.

The book should also be useful for students on degree courses overscas, where instruction is in English, because, although it uses CP 110 in applications, the principles apply to large parts of all degree courses. In many other countries the students have not had the benefit of similar books in their own language to help them in the past, but have had to depend solely on their Professor's notes. These students are becoming more conversant with English so this book might fill a long awaited need to help them in their professional/ degree courses.

Additionally, the book is considered to be explicit enough, and covers all the work concerned with concrete for HND, HNC, and architectural students.

One problem has been space, as just reproducing CP 110 would more than fill the book, and each one of the topics, properties of concrete, reinforced concrete, shells and folded plates, and pre-stressed concrete, might be thought to need a text book of the length of this book. To adhere to the requirements of students, with each topic the writer has avoided giving potted histories (e.g. it is not necessary to know some small pieces of Roman engineering to understand CP 110) and research treatises, but has concentrated on presenting the technology as it is now in as lucid and useful a way as possible. Yet relevant research is either known or has been reviewed and has had its effect on the work presented in this book. Examples to illustrate trivial points are omitted. Also, for economy the book avoids introducing and briefly reviewing a topic, only to refer the reader elsewhere for a proper study of the topic. If a topic is not included because of its priority rating with regard to a student's bachelor degree course, or the needs of a designer, relative to its demands on space, then it is usually completely omitted and the reader needs to make use of a library, or his lecturer, for that particular topic.

References have been kept to a minimum and the ones best known to the writer, and he apologises to numerous academics/ researchers for excluding their meritorious works. Students can usually obtain text books from libraries more easily than past, particularly very old, research papers, so the writer often refers to recent text books rather than to the original contributions and apologises to many eminent original researchers. References for further reading have not been given as, because of the explosion of literature in this field, choice becomes subjective. They can easily be obtained from various libraries of specialist research associations and professional institutions. For clarity of writing, repetition rather than cross-referencing has been used for small points. If a reader finds he does not know about a topic mentioned in a sentence then he can refer to the index.

Past examination questions have been excluded as these tend to be obsolete because of CP 110. Examples illustrating each non-trivial topic in the book will tend to form the basis of future examination questions. Where CP 110 is used the reader is assumed to have CP 110 available. In the examples, except for shells and folded plates, slide rule accuracy is usually adequate although greater accuracy is often given because it was available from an electronic calculator.

Chapter 1 endeavours to introduce concisely the concepts of CP 110 with regard to limit state, characteristic strength and γ factors.

Chapter 2 gives more than generally required by bachelor degree courses, with regard to the types and properties of cements and concretes, including the present method of mix design, which is also useful to many practising engineers. It also gives information on reinforcement and its anchorage, and gives design tables useful to students and practising engineers for quickly determining anchorage lengths in accordance with CP 110.

Chapters 3 and 5 give the theories which are the basis of CP 110 for the various limit states of design of reinforced concrete beams and columns. Examples of the use of CP 110 and design tables are given for the use of students and practising engineers.

Chapter 4 gives the methods of designing various slabs and includes the methods of yield line (Johansen) and Hillerborg's strip. For both these methods examples are given which should help the beginner immensely, since the methods appear very formidable when one is confronted with the many existing publications, which are difficult and time consuming for students to follow.

Chapter 6 explains how the plastic theory of design, commonly used for indeterminate steelwork structures, can be adapted to the design of reinforced concrete structures.

Chapter 7 indicates how complete structures are designed. It gives information on the creation and choice of the most suitable structure. Most books on design tend not to give this most valuable information but perform designs of already created structures. Exercises are suggested for students in their design offices. Again, the difficult task of explaining to the inexperienced which type of construction to use is tackled more generally in Section 2.5, even though bold statements have to be risked.

Chapter 8 gives the principles required for designing prestressed concrete according to CP 110. For example, the principles of, say, a certain loss are described, then as regards quantifying it for a designer to use, for economy of space, just the CP 110 value is considered and not the numerous alternatives. For overseas engineers the principles and application are useful but the quantifying has to be extracted from their own Codes. The writer uses this chapter for teaching bachelor and master degree students at the University of Bradford.

Chapter 9 deals with shell roofs and folded plates and is used in teaching MSc students in structural engineering at the University of Bradford. In this part-time course the students learn or improve their reinforced concrete detailing by using the conoidal shell tables at the beginning of the course. Many of the students are poor at reinforced concrete detailing, yet they will have detailed simple beams as undergraduates and/or practising engineers (not employed with specialist reinforced concrete firms), but generally none will have detailed a conoidal shell roof before and by using the tables they are not handicapped by a formidable design time before they can detail under supervision. The writer finds that students are understandably slow at design and if a design is intended to train them at detailing reinforcement they sometimes spend all the periods de-designing and then make the drawings afterwards without proper supervision. The design tables enable them to detail in their drawing office periods under supervision. Similarly, design tables for folded plates have been given for the same purpose. These tables could also be used for undergraduates. Conoids and folded plates have the advantage of being sophisticated structures which students will most likely not have the opportunity to experience in practice, hence they are more intellectually satisfying than simple beams for learning detailing in a university. All tables in this book have been prepared with every care, but are not intended to be used to take away from a consultant responsibility for his own designs. He may use the tables, but should assure himself, by checking perhaps with other tables, computer programs, or his own calculations, or at the very least by calculations equating internal stresses and external forces and moments, that his design is sound. Alternatively he may prefer to use the tables as a rapid check on his calculations.

The notation is defined where it is used and is in accordance with the international notation. The writer used the latter for shells but eventually found it unsatisfactory, concluding that the international notation for concrete is not relevant to shell analysis.

The writer has been able to have the benefit of CP 110 being finalised and in use for some time, and has benefited from being able to study, before completion of this book, the following British books based on CP 110:

Hughes, B. P., *Limit State Theory for Reinforced Concrete*, Pitman (1971).
Regan, P. F., and Yu, C. W., *Limit State Design of Structural Concrete*, Chatto (1973).
Bennett, E. W., *Structural Concrete Elements*, Chapman and Hall (1973).
Allen, A. H., *Reinforced Concrete Design to CP 110*, C. and C. A. (1975).
Astill, A. W., and Martin, L. H., *Elementary Structural Design in Concrete to CP 110*, E. Arnold (1975).
Kong, F. K., and Evans, R. H., *Reinforced and Prestressed Concrete*, Nelson (1975).

In the case of properties of concrete, the 1973 edition of A. M. Neville's *Properties of Concrete* (Pitman) was also of help. The writer has also benefited from being able to study the U.S.A. editions of:

Winter, G, and Nilson, A. H., *Design of Concrete Structures*, McGraw Hill (1972).
Fergusson, P. M., *Reinforced Concrete Fundamentals*, John Wiley (1973).

The following approximate figures may be helpful to British, American, Canadian, etc., readers:

Imperial	U.S.A.	SI (approx.)	SI (accurate)
1 ton	1 long ton	10 kN	9964 N
2000 lb	1 short ton	9 kN	8896 N
1000 lb/in²	1 kip/in²	7 N/mm² = 7000 kN/m²	6895 kN/m²
1 lb	1 lb	4.5 N	4.448 N
1 in	1 in	25 mm (100/4 mm)	25.40 mm
1 ft	1 ft	0.3 m or 300 mm	304.8 mm
62.4 lb/ft³	62.4 lb/ft³	10 kN/m³	9802 N/m³
1 ton/ft²	1 long ton/ft²	107 (say 110) kN/m²	107.3 kN/m²
1 ton/in²	1 long ton/in²	15.5 N/mm²	15445 kN/m²
20 lb/ft²	20 lb/ft²	1 kN/m²	957.6 N/m²
150 lb/ft³	150 lb/ft³	24 kN/m³	23.56 kN/m³
1000 lb/ft	1000 lb/ft	14 or 15 kN/m	14.59 kN/m
3 ton/ft	3 long ton/ft	100 kN/m	98.07 kN/m

The following approximate figures might be useful for readers in metric countries not yet accustomed to SI units: $10 \text{ kg/cm}^2 = 1 \text{ N/mm}^2$, and 1 tonne = 1000 kg = 10 kN. (Note that $g = 9.807$ m/s² and the terms *force* and *mass* have not been used above.)

The writer has been impressed by certain firms managing very well with SI units by concentrating on kN and metres as units for calculations, and has followed this practice in this book. We used to desire larger units than pounds and inches and sometimes used kips and feet (for structural analyses). The kN and metre are thus useful sized units for calculations and avoid large powers of 10 required when using N and mm units.

Acknowledgements are due to several firms which specialise in reinforced and prestressed concrete and which have employed the writer in the past; also to the Universities of Leeds (Prof. R. H. Evans), Sheffield (Prof. N. S. Boulton), and Bradford for research facilities. I would also like to acknowledge certain technological help from Prof. A. L. L. Baker, formerly Professor of Concrete Technology, Imperial College, University of London, Dr A. W. Beeby of the Research and Development Division of the Cement and Concrete Association, Dr E. W. Bennett, Reader in Civil Engineering, University of Leeds, and Dr Christian and Mr Franklin of the University of Bradford.

The writer has found discussions with his sons, C. Anthony Wilby and Chris. B. Wilby, useful with regard to the styles of books and points liked by students of civil and structural engineering, i.e. a consumer's viewpoint.

C. B. W.

Contents

1 **Limit State Design and CP 110**   1
2 **Properties of Materials**   3
3 **Reinforced Concrete Beams**   58
4 **Reinforced Concrete Slabs**   100
5 **Columns and Walls**   111
6 **Reinforced Concrete Frames and Continuous Beams and Slabs** 121
7 **Design Using CP 110**   124
8 **Prestressed Reinforced Concrete**   141
9 **Shell and Folded Plate Roofs**   172
  Index   209

Tables and graphs for designers

Units. Conversion of British Imperial and U.S.A. into SI units   ix

Mix design
   Graphs plotting percentage passing against sieve aperture size for concrete
     aggregate   10
   Table recommending suitable workabilities for various uses   13
   Graphs plotting average ultimate compressive stress against water-to-
     cement ratio for concretes of various ages   15
   Table recommending minimum strength as percentage of average strength
     for various conditions of control of concreting   20
   Tables recommending aggregate-to-cement ratios for various gradings and
     types of aggregates, water-to-cement ratios, and workabilities   24, 25
   Table showing how to determine a certain required grading from available
     sand and coarse aggregates   27

Weights of materials. Weights of materials in kN/m^3 and kN/m^2 139

Reinforcement
   Table giving cross-sectional areas of numbers of bars and bars in slabs   69
   Tables giving CP 110 values of f_y for various types of reinforcement bars   43

Anchorage or bond lengths
   Table giving tension anchorage lengths (l_b/d_b) for various values of f_{cu} and f_y 43
   Table giving compression anchorage lengths (l_b/d_b) for various values of
   f_{cu} and f_y 44
   Table giving anchorage length equivalents of hooks and nibs for various
     diameters of mild and high yield steel bars   46
   Table giving compression and tension anchorage lengths for $f_{cu} = 20\,N/mm^2$
     and various values of f_y 47
   Table giving overall anchorage lengths using hooks and nibs for $f_y =$
     $250\,N/mm^2$ and $f_c = 20\,N/mm^2$ 47

Table giving tension anchorage lengths for bars used in water-retaining structures, CP 2007   56

Curtailment of bars in beams. Table giving points for stopping off or bending up tension reinforcement bars towards supports for simply supported, continuous and fixed beams   48

Elastic theory
Table giving tension anchorage lengths for bars used in water-retaining structures, CP 2007   56
Tables for calculating equivalent area, x and I 65, 71
Tables giving corresponding values of $K [= M/(bd^2)]$ and z_1 for $f_s = 85\,\text{N/mm}^2$ and $\alpha_e = 15$ 70

Shear reinforcement
Table giving values of V/d for two-arm stirrups for various values of f_{yv}, d_b and s_v 80
Table giving values of V for single bars bent-up at 45° in single shear for various values of f_{yv} and d_b 81

Plastic design of sections for bending moments. Table giving, for balanced design, values of $K_1 [= M_u/(bd^2)]$ and $\rho\% [= 100A_s/(bd)]$ for various values of f_y, f_s and f_{cu} 91

Strength of steel in compression
Table giving f_{sc} (design ultimate compressive stress) for various values of f_y for compression steel   95
Figure giving CP 110, Fig. 3, stress–strain curve for 25 mm diameter alloy bar   161

Design of beams and slabs. Table to assist in the preliminary design of depths of beams and slabs of various spans   126

Continuous beams and slabs
Tables giving bending moments and shear forces in continuous beams and slabs carrying dead and imposed loadings   123
Table giving bending moments and shear forces in continuous beams and slabs subjected to unit bending moment at one and both ends   138

Single span beams with fixed and free end supports. Table giving bending moments, support reactions and deflections for beams with various loadings   137

Cylindrical shell roofs
Graphs for estimating reinforcement in schemes of normal and North-Light shell roofs   178
Elastic analysis using tables for designers   186–188

Conoidal shell roofs. Elastic analysis using tables for designers   188–191

Folded plate roofs
Tables for designing folded plates of any practical geometry   193–200
Tables for designing certain practical schemes of folded plates   201–207

Chapter 1

Limit state design and CP 110

1.1 Limit state design

A structure must be safe and serviceable. Safe against collapse means that the ultimate strength of a member must equal the *load factor* times the strength required of the member to resist *working loads*. Load factors are generally about 1.8. Working loads are the loads which one predicts that the member will normally be required to carry. Serviceable means that the member must be satisfactory at working loads. Normally this means satisfactory as regards (a) deflection—this must be limited so as not to cause damage to partitions and finishes nor undesirable appearance nor inefficiency (for example people feel insecure if walking on a staircase or floor which noticeably oscillates when in use), and (b) cracking—this should not adversely affect the appearance or durability of a member. One can think of many more serviceability requirements, e.g. members should be satisfactory at working loads with regard to durability and resistance to vibrations, fatigue, fire and lightning, etc.

In a design all the above have to be limited—thus the term *limit state design*, and CP 110 uses the expressions *ultimate limit state* (for safety) and *serviceability limit state* (for serviceability). The latter comprises *serviceability limit state of deflection*, etc.

Previously (e.g. CP 114) stresses, called *working stresses*, were determined at working loads with an elastic analysis. These were then restricted to *permissible stresses* for the concrete and steel. The permissible stress of a certain material was equal to its ultimate strength (stress) times its *factor of safety*. These were chosen by experience so that members designed in this fashion were adequately safe, and free of troublesome cracking at working loads. The elastic theory enabled a designer to calculate deflections at working loads to check that they were satisfactory. This method was very satisfactory over many years. Its main weakness is using an elastic theory to satisfy safety. When one wishes to take full advantage of stronger modern materials one must use a plastic theory for satisfying safety requirements. The elastic theory of design tends to be unduly

1

conservative for predicting ultimate strength. However, it is still used for water retaining structures (CP 2007). With unlined or inadequately lined tanks, permissible concrete and steel stresses are kept low so that cracking does not occur. It will of course occur to some extent due to shrinkage, which is ignored by CP 2007 requirements. However, we know from experience that the design restricts cracking adequately for watertightness. The elastic theory is a fairly good predictor of stresses at working loads. When a tank is designed with these low permissible stresses inevitably its safety will be very conservative.

1.2 Characteristic strengths (stresses) and loads

Characteristic strengths are explained using concrete cubes as an example in Section 2.3.9. The same applies to reinforcement and prestressing tendons. The scatter of the results (the standard deviation) would be much less for those than for concrete cubes. As for cubes, the characteristic strength of steel is the value, of the yield or proof stress of reinforcement or the ultimate load of a prestressing tendon, below which not more than 5% of the test results fall.

Characteristic loads are based on the same considerations.

1.3 γ_m factors of CP 110

It is considered that the strength of the materials in the actual structure will depart to some extent from the strengths of control cubes and steel specimens. Hence for design, the characteristic strength is divided by a factor γ_m; the result is called a *design strength*. As one would expect γ_m is greater for concrete than steel. Many factors can affect concrete strength on the site after cubes have been taken, whereas steel suffers less (a small surface rusting say). For concrete γ_m is normally 1.5, but 1.3 for considerations of localised damage or excessive loads. The corresponding γ_m figures for steel are 1.3 and 1.0 respectively.

Chapter 2

Properties of materials

2.1 Cement

Cement is the most important and expensive ingredient of concrete. It was patented by J. Aspdin in the U.K. in 1824 and he called his product *Portland cement* because the artificial stone (concrete) made with it resembled Portland stone.

Portland cement is made by grinding together its principal raw materials, which are (a) argillaceous, e.g. silicates of alumina in the form of clays and shales, and (b) calcareous, e.g. calcium carbonate in the form of limestone, chalk, and marl which is a mixture of clay and calcium carbonate. The mixture is then burned in a rotary kiln at a temperature between 1400 and 1500°C; pulverised coal, gas or oil is the fuel. The material partially fuses into a clinker which is taken from the kilns, cooled and then passed on to ball mills where gypsum is added and it is ground to the requisite fineness. The resulting cement is allowed to have certain maximum percentages of materials not required, some disadvantageous for some uses, such as iron oxide and sulphur trioxide. A general idea of the composition of cement is indicated by the following oxide composition limits of Portland cements: lime (CaO) 60–67%, silica (SiO_2) 17–25%, alumina (Al_2O_3) 3–8%, iron oxide (Fe_2O_3) 0.5–6%, magnesia (MgO) 0.1–4%, sulphur trioxide (SO_3) 1–3%, soda (Na_2O) and/or potash (K_2O) 0.5–1.3%.

The constituents forming the raw materials used in the manufacture of Portland cement combine to form compounds, sometimes called Bogue[1] compounds, in the finished product. The following four compounds are regarded as the major constituents of cement: tricalcium silicate ($3CaO.SiO_2$ or C_3S), dicalcium silicate ($2CaO.SiO_2$ or C_2S), tricalcium aluminate ($3CaO.Al_2O_3$ or C_3A) and tetracalcium aluminoferrite ($4CaO.Al_2O_3.Fe_2O_3$ or C_4AF).

A cement works is usually sited near to its raw materials. These sites vary and consequently cements from different works vary within permissible limitations. In the U.K. this variation seems to have an insignificant effect upon concrete. However, research

3

by the author and others indicates that the asbestos cement manufacturing process is sensitive to the percentage of C_3S, which varies significantly with cements from different works in the U.K.

High alumina cement was first made by J. Bied for the French Lafarge Company in 1908, and named *Ciment Fondu*. This discovery was made whilst searching for a cement which liberated no free hydrated lime upon setting. Portland cement liberates free hydrated lime upon hydration and this in the resulting concrete is very vulnerable to attack from mineral sulphates, dilute acids and other agents.

When cement is hydrated, lime and alumina are liberated. The lime combines with the alumina and in the case of Portland cement an excess of lime results, whereas in the case of high alumina cement an excess of alumina results. Bearing this in mind, the properties of these two fundamentally different cements can often be predicted. For example, when these cements are mixed together and hydrated, the respective excesses of lime and alumina react chemically with one another and a *flash set* (almost instantaneous setting) can result. This can be useful for caulking small leakages in cofferdams and water retaining structures. The flash set phenomenon is, however, a reason for new *Ciment Fondu* concrete not being suitable for jointing to new Portland cement concrete, and vice versa. Time limits have to elapse so that there is no danger of unhydrated Portland cement coming into contact with unhydrated high alumina cement. The concrete which is to be extended should be 24 h old if it is *Ciment Fondu* concrete, 2 days old if rapid hardening Portland cement, and 7 days old if ordinary Portland cement.

When cement is hydrated the terms *initial setting time, final setting time* and *rate of hardening* are used, often loosely. However, the first two are defined for cement by BS 12, 915 and 1370. Other tests of cement for soundness, tensile and compressive strength, chemical composition, fineness of grinding, etc., are described in BS 12. The definitions of *initial set* and *final set* unfortunately bear no precise relationship to practice. They do, however, enable the properties of different cements to be compared for their setting qualities. It can loosely be said that it is good practice not to disturb concrete after its initial set, and the initial setting time is normally not less than half an hour. There are exceptions to this rule in practice, however, since such operations as the trowelling of concrete floors and granolithic finishes, for example, usually need to be performed after the initial set, but before the final set has taken place. The final setting time is not usually more than ten hours.

If one imagines say a sown-up sheep's bladder (a colloidal membrane) containing a solution, immersed in a similar solution of greater dilution, then water travels through the very fine pores in the bladder so that a pressure (osmotic) is developed in the bladder. This pressure continues to increase until the solutions on either side of the colloidal membrane have the same dilution. This is a very simple description of colloidal chemistry relative to the hydration of cement. Upon hydration the surface of a small portion of cement forms crystalline substances, which can be observed with an electron microscope[2], with the water. These form a colloidal membrane, surrounding the portion of cement, called *tobermorite gel*[3] (a calcium silicate hydrate). As indicated previously, water travels through the membrane to dilute the solution of hydrating cement compounds within the membrane. This causes a pressure inside the membrane and hence expansion of the concrete or mortar. Conversely, drying of the cement after hydration causes *shrinkage* of the concrete. However, the amount of shrinkage caused by complete drying out of hydrated cement paste is not completely recovered by subsequent wetting.

If water is in contact with concrete, e.g. the wall of a basement, water can travel through the concrete not only via any cracks, construction joints, or voids, but also through the colloidal membranes in contact with it, the water then passing through the adjacent colloidal membranes, until all solutions surrounded by colloidal membranes have reached the same dilution. If the inside of the basement is at a lowish humidity, say, then this will take water through the colloidal membranes in contact with this air. Thus water, or dampness, can be transmitted through a basement wall, hence the desirability of 'tanking' even if the concrete is very good.

The strength of a cement paste depends greatly upon the bonds formed between the very small particles of its cement gel. Generally the greater the number of these particles and the denser the gel structure, the stronger the gel mass[2]. Hence the water-to-cement ratio used for a cement paste is related to its strength.

There are several types of cement available to the engineer, for example, as follows:

1. *Ordinary Portland cement*. This is the most inexpensive cement and is consequently widely used.
2. *Rapid hardening Portland cement*. As the name implies, concrete made with this cement hardens more rapidly than concrete made with ordinary Portland cement. Such a property enables early stripping of concrete formwork, especially advantageous for precast work where repeated uses are made of the same shutter. *Extra rapid hardening* cements can be obtained for special purposes.

These two cements are of the same material as 1 except more finely ground.

3. *High alumina cement.* This cement is not classed as a Portland cement. It hardens much more rapidly than any other commercial cement, and it has the further advantage of being sufficiently immune, for practical purposes, to attack from several important chemicals. Some examples are: many of the sulphates present in subsoil waters and in sewage; sulphur compounds formed from the combustion of coal and oil; carbonic acid as experienced in subsoil waters from moorland areas; many of the chemicals contained in sea water; chemicals which attack Portland cement and which are present in important industries such as lactic acid (associated with milk), tar oil, cottonseed oil, beer, and sugar juices. *Ciment Fondu* was excluded from CP 110 by the August, 1974, amendment, but was previously allowed to be used when high strength was required urgently, for example on maritime structures when it was necessary to have a reasonably hard concrete before high tide; for the sealing of water leaks in emergencies when excavating in water bearing ground; for structural work which required to be in use within say 24 h; for structural work where formwork was required to be stripped early or where it was required to prop further shutters from the members cast as soon as possible; for prestressed concrete, especially pretensioned concrete, where economy required release of the wires and removal of the members from the stressing beds as early as the strength of the concrete permitted. The high early strength is obtained to some extent because the chemical reaction of the cement with water is very exothermic. To avoid the ills of overheating (see 7, page 8) it is desirable to have a low water-to-cement ratio (to reduce the rate of chemical activity), to cast at an ambient temperature of not more than about 20°C, not to allow the internal temperature of the concrete to be more than 30°C for more than 24 h after casting, to cure with water or similar, and certainly not to steam cure.

The greatest disadvantage of high alumina cement was its cost, which made it prohibitive for many purposes. Another economic disadvantage was the necessity of curing with water or dampness. Concrete using this cement was nevertheless quoted as being more economical than steam cured Portland cement concrete for prestressed concrete work.

Ciment Fondu with a suitable aggregate can be used as a refractory concrete or mortar for fireclay bricks and is suitable for temperatures up to about 1300°C. High climatic temperatures in combination with high humidities as experienced in the tropics were found to reduce the strength of concrete made with *Ciment Fondu* rather alarmingly[4]. The chemical conversion of certain crystalline compounds having certain numbers of elements of water of crystallisation to other crystalline compounds with different numbers of elements of water of crystallisation could cause an internal volume change in the concrete with a consequent disruption and weakening of the concrete. The shape of crystals changes from hexagonal to cubic. Neville[4] claimed that this chemical conversion could also eventually occur with aging in the cool damp U.K. climate, although CP 110 prior to the August, 1974 amendment, regarded this effect as negligible for properly cured concrete. It might be thought that high alumina cement concrete could be used in structures protected from moisture, which is the case with many buildings, without worrying about chemical conversion. At the time of writing, however, there is considerable worry because of certain failures in the U.K., and even the latter type of structure is receiving investigation. Even indoors, with central heating and solar gain through large glass windows, temperatures can be high and it is argued that there is always water in some form inside the concrete, and the humidity of the atmosphere can be high in the U.K. and this air is not normally dried before entering buildings. After full conversion, concrete strength increases with age.

Although the dangers of conversion became rather catastrophically experienced about 1961, seemingly inadequate notice had been taken of this since then, until about 1974 when there was considerable alarm concerning lack of reliable knowledge of when high alumina cement could be used. Inadequate notice was taken of work by Bolomey[5] of France in 1927 and Davey[5] in the U.K. in 1937; both demonstrated that high alumina cement concretes, hardened under good conditions, subsequently lost up to 40% of their strength permanently, due to curing in warm water, and experienced the colour change to yellow-brown, which we now know to be due to conversion.

In the case of the most publicised failure in the U.K., the pre-stressed concrete beams were over a swimming pool and experienced warmth, moisture from condensation and roof leaks, sulphate attack from the plaster, and possibly had poor concrete and support seatings. The other few failures seem to have had more than just conversion as a weakness. Subsequently most high alumina cement work has been tested in the U.K. and most of it found to be safe. The author has tested roofs in buildings with up to 95% conversion and found them very safe. There is no doubt that steam constantly directed on to high alumina cement beams can cause disintegration.

4. *Cement for use in cold weather*. Such cements are usually achieved by adding about 1.5% of calcium chloride to rapid hardening Portland cement. The calcium chloride generates heat by reacting

with the water used in mixing the concrete. This also enhances the rapid hardening qualities. Because of the heat evolved, these cements can very often be profitably used in cold weather to allow concreting operations to continue. The high early strength properties are advantageous for allowing early stripping, and, in the case of precast concrete, handling. The chloride ion aggravates the corrosion of steel (this is particularly so in the case of NaCl). Hence if water and oxygen ion can penetrate to the reinforcement through pores and/or cracks in the concrete, the calcium chloride will increase the rate of corrosion of this reinforcement. It is interesting that in the case of water retaining structures and underground pipelines, if the water is in contact with concrete containing very fine cracks which penetrate to the reinforcement, it is possible for corrosion to occur even though many would not imagine that air could penetrate through the crack. This is because the oxygen ion of air dissolved in the water is easily carried in the water penetrating the crack to the steel— refer to the theory of notch corrosion. CP 110 prohibits calcium chloride in prestressed pretensioned concrete, and restricts it to not more than 1.5 % by weight of the cement in reinforced concrete.

5. *Sulphate resisting cement.* This cement is made specifically to resist the attack of sulphates. Underground structures can experience sulphate attack from the soil, back-fill or ground water. There is a cement known as *super sulphated cement* which resists greater concentrations of sulphates.

6. *Cements with a low coefficient of shrinkage* can be specifically devised for highways, dams, water retaining structures, etc., to reduce the magnitude of cracks caused by shrinkage. Such a cement, which also had low heat of setting, was devised and used for the mass concreting to the Boulder Dam, U.S.A. There are cements which claim to expand, but they do not always do so if the concrete subsequently dries out.

7. *'Low heat' Portland cements* generate less heat upon reacting with water than normally experienced with other cements and are thus suitable for mass concrete work. The heat generated with Portland cement in mass concrete work can literally boil off the water required for the necessary chemical reaction, the steam causing flash setting of some of the cement and also disruption and voids in the resulting concrete.

8. *'Portland-pozzolana cements'.* Fly ash, pulverised fuel ash, or pozzolana is sometimes substituted for 10 % by mass of the ordinary Portland cement to achieve low heat of setting and reduced shrinkage without reducing the 28 day strength of the concrete, but the early rate of hardening is reduced. Recently this idea has been used for a gravity dam in Yorkshire, and to help further, the concrete mix has

a low cement content and large size aggregate. Unfortunately fly ash contains a small amount of sulphate and CP 110 restricts the total sulphate content of a mix expressed as SO_3 to not more than 4 % by mass of the cement.

9. *Coloured cements* are used for reconstructed stones, renderings, and the like. Because of the high cost of these cements, coloured artificial stones usually have a facing about 38 mm thick made with the coloured cement, and a backing made with ordinary Portland cement. Coloured cements can be obtained by adding the following pigments to Portland cement: yellow ochre (yellow), brown oxide of iron (brown), green oxide of chromium (green), red oxide of iron (red), manganese black (black). The weight of the pigment should not exceed 10 % of the weight of the cement, otherwise the strength will be impaired. White cements are popular and require to be specially manufactured. The colour of a concrete can be improved and will wear better if the aggregates are also of a similar colour to the cement.

10. *Portland blast furnace cement* is obtained by grinding granulated blast furnace slag with the clinker which is normally ground down to make ordinary Portland cement. It has a slightly lower heat of hydration than ordinary Portland cement, is slightly more resistant to sulphate attack, and is slower to develop its early strength.

11. *Water-repellant cements.* Certain ones are most effective in sealing leakages in water retaining structures.

2.2 Aggregates

Aggregates are classed as *fine aggregates* and *coarse aggregates.* Generally, various sands are used as fine aggregates, and coarse aggregates are either water worn gravels or crushed rocks. The aggregates chosen are usually the most inexpensive to give the requisite quality of concrete. The engineer must however be satisfied that the source selected will consistently supply the quality of aggregate which he has approved. Sometimes the engineer requires stockpiles at the suppliers' works to meet with his approval. These are then drawn upon exclusively for the concreting operations.

Aggregates for normal concreting work are a fairly inexpensive commodity at the quarry and thus transport charges substantially influence their overall cost. Local aggregates are therefore generally employed, but an expensive type of aggregate may warrant greater transport costs if the necessary stone does not occur locally. Examples of more expensive stones are: granites for granolithic finishes; various types of coloured aggregates for artificial (reconstructed) stones (usually used for the surface layer of the stone only); and vermiculite for lightweight finishes (imported into the U.K.).

Reference should be made to the British Standards 882, 1198, 1199, 1200 and 1201, which recommend various gradings of the particle sizes for both fine and coarse aggregates. These enable standardisation and control but are not necessarily ideal gradings for concrete. The standards quoted specify tests of other relevant qualities of the aggregates, namely specific gravity, water absorption, bulk density,

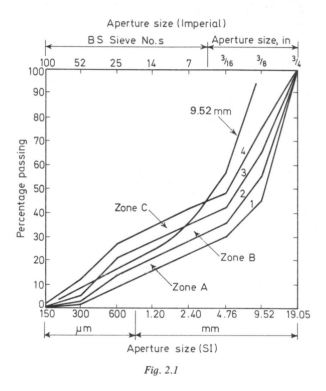

Fig. 2.1

organic impurities, and crushing strength. *Figure 2.1* shows four gradings, upon which the mix designs of *Road Note No. 4* are based, for 19.05 mm ($\frac{3}{4}$ in) and down aggregates, and one average grading curve for 9.52 mm ($\frac{3}{8}$ in) aggregate. The grading of a 19.05 mm ($\frac{3}{4}$ in) aggregate should lie within curves 1 and 4 and preferably within curves 2 and 3 if this method of mix design is to be used.

Coarse aggregates can be classified according to shape (BS 812) as follows:

1. *Rounded aggregates*, for example beach and other well worn gravels.

2. *Irregular aggregates*, for example water worn river gravels.

3. *Angular aggregates*, for example crushed rock or manufactured materials. These are commonly granites, limestones, basalts, quartzites, flints, pumice, broken bricks, foamed slag, blast furnace slag, sometimes a strong sandstone, vermiculite and duromit, etc.

The grading, shape, porosity and surface texture of the aggregates can affect the workability and consequently the strength of concrete.

When a concrete is required to be lightweight, to have a good resistance to heat transmission and impermeability to water, and a high strength is not required, special lightweight aggregates are often used, such as vermiculite, foamed slag, clinker, breeze, pumice, wood wool, expanded shales, etc.

If water is added to 1 m³ of sand, the gross volume of this sand increases until it occupies about 1.25 m³. After this volume is attained the addition of further water decreases the gross or bulk volume until when the sand is finally saturated the volume has returned to 1 m³. When concrete is 'batched' by volume (i.e. the ingredients measured by volume) the water content of the sand greatly influences the quality of the resulting concrete. Consider a 1(cement): 2(sand): 4(gravel) mix, the ratios referring to dry volumes of the respective materials (as is standard practice). If we were using a sand experiencing its maximum amount of 'bulking' of say 25%, then the mix actually produced in terms of dry volumes would be $1:2/1\frac{1}{4}:4$ or $1:1.6:4$.

If water is added to 1 kg of sand, the gross weight is increased by the weight of the water added to about 1.1 kg upon saturation. Hence, if the batching of concrete were by weight, the water content of the sand would still be troublesome but not to as great an extent as by volume. Consider again a 1:2:4 mix and let the sand be increased in weight by its maximum amount of say 10% due to its water content. Then the mix actually produced in terms of dry volumes could be $1:2/1.1:4$ or $1:1.818:4$. For illustrative purposes it has been assumed that the bulk densities of the dry materials are the same. Thus the inaccuracy of batching by weight is basically not as great as batching by volume. This reasoning ignores the fact that the same phenomenon also affects coarse aggregates, but to a far lesser extent. Several devices are available for measuring the water content of the aggregates, so that the mix can be adjusted accordingly. The water content often varies from place to place in a stockpile. When a large concreting programme is being conducted, sometimes the stockpiles will be insufficient (especially on congested sites) and sand which arrives during the course of the concreting operations will have a different water content to the sand in stock. Aggregates are commonly exposed to the weather so that the water content will vary with the rainfall. One needs to be vigilant therefore to allow for the errors in batching caused by the water content of the aggregates.

2.3 Concrete

Coarse aggregate, fine aggregate, cement and water are mixed together in suitable proportions, and this mixture, placed and compacted wherever required, solidifies after a lapse of time into what is known as concrete.

The mixes of concrete commonly used for structural purposes are 1 part (by dry volume) of cement:2 parts (by dry volume) of fine aggregate:4 parts (by dry volume) of coarse aggregate, and similarly, $1:1\frac{1}{2}:3$ and $1:1:2$.

Many investigators have proved that most of the qualities desired of concrete benefit by increased compressive crushing strength, for example, strength in tension, shear, resistance to weathering, abrasion and wear, and impermeability. Exceptions to this rule are lightness (in density), thermal insulation and shrinkage.

The factors which have the greatest effect upon the strength of concrete are the cement-to-aggregate ratio, the compaction, the water-to-cement ratio of the mix, and the method of curing.

It is easy to imagine that the strength of concrete depends upon the absence of voids, or in other words, upon the final density after setting and maturing. For example, 5% of air voids can give a loss in strength of 30%, 10% of voids can give a loss in strength of 60% and 25% of voids can give a loss in strength of 90%. Compaction of the concrete is therefore extremely important, and this is dependent upon the workability of the concrete.

2.3.1 Workability

Workability is the ease with which concrete can be placed in moulds, compacted around reinforcement and screeded to a level. Many tests have been devised for measuring this property, and all have been subjected to much adverse criticism. The test which has possibly been condemned the most, namely the *slump test*, is the most commonly used in the U.K., and is referred to by CP 110. The nature and the grading of the aggregates considerably affect the slump. Thus specifying the slump can ensure uniformity in the consistency of concrete during the progress of work only if the materials are of constant quality.

Other tests of workability referred to by CP 110 are the *compacting factor test* and the *VB consistometer test*. The former was developed as an improvement upon the slump test in attempting to measure workability. The latter became useful in the U.K. when drier concretes than previously became necessary for prestressed concrete work, as it can distinguish between various concretes having virtually zero slump. It is also better for very dry mixes than the compacting factor test.

Table 2.1.

Degree of workability	Slump, mm	Compacting factor		Use for which concrete is suitable
		Small apparatus	Large apparatus	
Very low	0–25	0.78	0.80	Vibrated concrete in roads or other large sections
Low	25–50	0.85	0.87	Mass concrete foundations without vibration. Simple reinforced sections with vibration
Medium	50 100	0.92	0.935	Normal reinforced work without vibration and heavily reinforced sections with vibration
High	100–180	0.95	0.96	Sections with congested reinforcement. Not normally suitable for vibration

Table 2.1 recommends suitable approximate workabilities of concrete for various uses.

Good compaction of the concrete, and hence a high strength concrete with a good finish, can be obtained by manipulation of the grading and type of the aggregates, the use of additives to reduce the surface tension of the water, employment of vibration and/or pressure, and use of a high water content.

The additives are *plasticisers* comprising soaps, detergents, or resins. Essentially they reduce the surface tension of the water, i.e. the water wets the particles more easily, increasing workability. They allow the water-to-cement ratio to be reduced for no decrease in workability, thus giving a stronger concrete. Some entrain finely dispersed air bubbles sufficiently for the concrete to have increased frost resistance for little decrease of strength—used for roads in cold countries.

The use of a high water content must be avoided as much as possible as it also decreases the strength of the concrete, as explained later. It can however be used with advantage when combined with a vacuum process (see page 16). Thus a high strength concrete requires to be as free from voids as possible. If water in excess of the amount required for the chemical reaction with the cement is present in the mix, this water remains in a free state and the concrete sets around the drops of water. Such particles of water eventually evaporate into the atmosphere, leaving pores and voids in the concrete, resulting in weakness and permeability.

2.3.2 Water-to-cement ratio and strength of concrete

The important effect of the water-to-cement ratio, by weights, on the strength of concrete was realised by D. Abrams of Chicago, in 1918, who stated that the strength of any workable concrete was dependent upon the water-to-cement ratio alone, assuming the same cement and degree of compaction are used and the conditions of curing and age at comparison of strengths are constants. The types of aggregates used can be varied, provided the concrete does not fail by the fracture of such aggregates. The workabilities of different mixes having the same water-to-cement ratios would be considerably different; for example a lean (low proportion of cement) mix might need vibration to obtain the same compaction as a richer (in cement) mix placed by hand. The strength of concrete increases as the water-to-cement ratio decreases, provided the water present is sufficient to allow the full chemical reaction to occur with the cement. If the water is less than this amount, a decrease in strength is experienced. *Figure 2.2* shows the relationship between the average ultimate compressive stress (or crushing strength) and the water-to-cement ratio for 150 mm cubes of fully compacted concrete for mixes of various proportions.

Only the compressive strength of concrete has been considered so far. It is generally accepted that this is a fairly reliable guide to the tensile and shear strengths, the modulus of rupture, the resistance to abrasion and wear, durability to the weather, density, porosity and watertightness. For durability, cement content is also important and minima are specified for various conditions in CP 110.

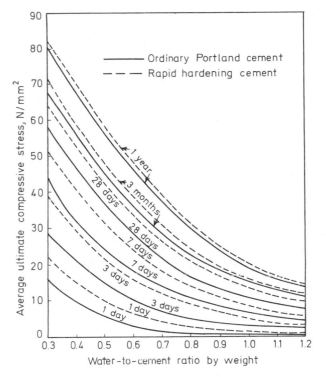

Fig. 2.2

2.3.3 Strength tests of concrete

BS 1881 specifies a standard compressive test, and also a standard test for the modulus of rupture. The latter flexural tensile test gives greater values than those obtained from tension tests made on standard briquettes (BS 12). The cross section of the briquette which is tested in tension is 25 mm square, the specimen being primarily designed for testing cements by determining the strengths of their cement:sand mortars. Larger specimens should be used for tension tests when the maximum size of the aggregate is greater than 9 mm. The cylinder splitting test has become popular as a tensile test of concrete. Unfortunately it is an indirect test of tension and assumes an elastic theory to calculate ultimate stress.

Shear in concrete beams is thought of in terms of diagonal tension and consequently the tensile strength of concrete is more relevant than the shearing strength. The shearing strength can be obtained from torsion tests of cylinders of concrete. The distribution of shear stress in such tests, however, is not the same as experienced in, say, a punching shear test.

With all the tests mentioned, size and shape of specimen matter, and thus empirical factors are usually required to relate these indicative control tests to the behaviour in the structural member.

2.3.4 Vacuum concrete

The concrete is made sufficiently wet to be placed and compacted easily and then the vacuum process removes water from the concrete, so that it finally has a low water-to-cement ratio. The water is extracted through mats placed in contact with the concrete. These mats are such that only water, and no cement, or fines (out of the aggregates) can be sucked from the concrete by the vacuum pump. Side shutters can usually be removed immediately afterwards if desired, as the concrete has almost zero slump.

2.3.5 Vibrated concrete

Concretes with low water-to-cement ratios can be placed and compacted by internal or external vibrators. External vibrators usually consist of motors with heavy cams on their shafts, and are fastened to a mould. Internal vibrators are of a poker type and can be held in the hand and immersed in the concrete where required. They are the more efficient for compaction and do not require the strong moulds often necessary for the external vibrators. If sufficiently dry mixes are used, the sides of the moulds can be removed immediately after vibration. There are in fact beam-making machines where the concrete is compacted by vibration, the sides removed immediately, and the beam dragged away along skids on its pallet. Most block-making machines employ pressure as well as vibration. Here again, solid and hollow blocks can be removed immediately from block-making machines on their pallets.

Workmen, when not strictly supervised, tend to make concrete extremely wet. Vibration does not increase the workability of such concrete and can be detrimental by causing segregation of the constituents of the concrete, the gravel tending to sink to the bottom, and the sand and cement to float to the top of the concrete. Such segregation can also occur with dry mixes if the vibration is sustained for a long enough period. The vibration employed with an apparently dry mix should be *only just sufficient* to make the concrete flow into the sharp arrises of the mould and around the reinforcement. Poker vibrators should not be removed rapidly or they can leave voids behind them.

2.3.6 Gap graded concrete

The principle of this method is to omit certain undesirable sizes of aggregates from the gradings, such as those of *Figure 2.1*. Undesirable sizes are those which prevent the efficient packing of the other sizes. If desired the smaller sizes of the coarse aggregate can be omitted, or one size only of aggregate can be used.

The more common aim of gap grading is to achieve strength from the efficient packing of the aggregate. This saves cement and allows aggregate suppliers to supply larger aggregate, less expensive to crush, which suits them also because there is a large demand for small aggregate for throwing with salt on winter roads in the U.K. By careful packing of stones, a strong wall can be built without using any cement. If a cement paste were to fill all the minor voids in such a wall, then a very strong construction would result, and this would be the ideal aimed at by the advocates of the gap grading of concrete.

A multitude of spheres of diameter D have a rhombohedral form of packing. These can be termed *major spheres*, and spheres of diameter $0.414D$, known as *major occupational spheres*, can fit into the voids between the major spheres. These spheres could, mathematically, constitute our coarse aggregate. The fine aggregate would then consist mathematically of *minor occupational spheres* of diameter $0.225D$, which would fit into the remaining voids. The voids now remaining can be fitted by *admittance spheres* of diameter $0.155D$, and these could also be provided by the fine aggregate. Cement would then occupy the remaining voids and a mathematically perfect compact mix would result. Such a mix, however, could not normally be cast in this ideal fashion and consequently some authorities[6] consider that only the major and admittance spheres are of practical value in designing a mix.

Mixes therefore are often designed with one size of coarse aggregate (e.g. 19 mm) and a sand, all the particles of which can pass through the voids in the compacted coarse aggregate. The sand is designed to fill the voids in the coarse aggregate and the cement is designed to fill all the remaining voids. The particles of sand must not be smaller than necessary, as this will increase the total surface area to be wetted with water and cement, and consequently a wetter mix (giving a weaker concrete) would be required for any requisite workability. Irrespective of the calculation just suggested, the sand should be sufficient to distribute itself uniformly throughout the mix under practical conditions. When the sand is less than 18 % of the mix it is difficult to obtain uniformity even under laboratory conditions. Mixes are often designed and then modified to suit the particular site conditions of mixing and compacting.

To increase workability it is advantageous to reduce the surface area of all the aggregates in a unit volume. This can be done by using larger particles. The largest aggregate possible should therefore be used, consistent with the minimum clearances allowed.

Gap grading enables leaner and drier mixes to be used, the absence of many intermediate sizes of aggregates having reduced the specific surface area of the aggregates and therefore having increased workability. The lean mixes usually utilised, however, make vibration almost essential. Such concrete, being made of leaner and drier mixes than a conventional concrete of equivalent strength, will therefore experience less shrinkage and hence possess better weathering qualities. Compressive forces on the gap graded concrete described are ideally transmitted from particle to particle of the coarse aggregate and not through any cement and sand particles. Consequently the creep associated with such concrete is low. A coarse aggregate as used in a conventional mix experiences a fair amount of segregation during transportation, and pouring into and out of lorries, etc. Rain also helps segregation in stockpiles. Gap grading avoids these disadvantages by requiring only single sizes of coarse aggregate.

Some advocate two different single sizes of coarse aggregates to be used with sand and cement in a mix. Gap graded concretes as lean as 1(cement):2.45(sand):6.59(gravel), with a water-to-cement ratio of 0.51, increase in strength with age in a similar fashion to conventional concretes[6]. Because of the packing of the aggregate of a gap graded concrete, vertical shutters can often be removed immediately after casting. Walls and columns can then be trowelled if desired or sprayed with a light water jet to expose the aggregate.

One disadvantage of gap grading is that if the single-size aggregates supplied contain over 2.5 % by weight of undesirable particles, this upsets the grading which is very sensitive to such intrusions. If however such irregularities are to be expected in the supply then the mix can be calculated accordingly to be of reduced efficiency.

2.3.7 No fines concrete

Coarse aggregate (gravel) is mixed with cement and the fine aggregate (sand) is omitted. *No fines* concrete is required to contain a multitude of voids to give good thermal insulation, and these voids need to be large enough to prevent the movement of water through the concrete by capillary attraction. *In situ* no fines concrete walls have been used in the U.K. for housing, the idea being that good thermal insulation is achieved and that rain beating on a wall only penetrates a short horizontal distance before having dropped to the bottom of the wall, there being no capillary paths to conduct the water completely through the wall. It is, however, often desirable to render and paint exposed walls.

2.3.8 Curing of concrete

After setting or solidifying, concrete increases in strength with age (see *Figure 2.3*). The strength at a particular age can be further

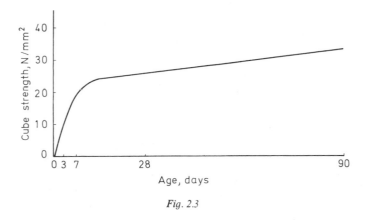

Fig. 2.3

increased by suitable curing of the concrete whilst it is maturing. Such curing comprises the application of heat (not if $CaCl_2$ is present or for high alumina cement or mass concrete) and/or the preservation of moisture within the concrete. The application of heat speeds up the chemical reaction and consequently rate of hardening of the concrete.

It can be imagined that preventing the escape of moisture from the concrete enables previously unwetted minute particles of cement to participate in the cementing action. If heat is applied to accelerate the hardening of the concrete it is therefore important not to expel the water held within the concrete. In other words, if heat is applied a high humidity is also desirable; steam is therefore a most suitable medium for this purpose. Steam curing can be done at atmospheric pressure or under pressure. The latter method is more effective but far more expensive, as pressure chambers are required. For example, the half-hour strength of concrete steam cured under pressure could equal the 28 day strength of an identical concrete maturing in air.

Increasing the strength of concrete by preventing the water used in mixing from escaping is usually done in one of the following ways:

1. *Flooding or submerging the concrete in water.* The floors of basements and reservoirs can fairly easily be flooded with water. Precast concrete units can be immersed in water in special tanks.

2. *Treating the surface of the concrete* so that it cannot dry out. Proprietary products exist for painting, or for applying coverings which adhere to the concrete.

3. *Covering the concrete with damp sand or hessian fabrics*, which are kept damp by watering periodically, or with thin polythene sheet.

2.3.9 Design of concrete mixes

Most commonly a concrete mix is designed to give the specified strength at the minimum cost. The cost depends upon the value of the materials, the labour required for batching, mixing, transporting, placing and trowelling, and the method of curing adopted.

Mix designs are fairly inaccurate due to the number of possible variables. The DSIR *Road Note No. 4* of 1950 based a mix design method on the aggregate gradings shown in *Figure 2.1*. This method is simple and can be used by mixing one's sand and gravel in such proportions as to correspond to one or other of these grading zones. As the method is even then not very accurate, it can be improved upon

Table 2.2.

Conditions	Minimum strength as percentage of average strength
Very good control with weight batching, moisture determinations on aggregates, etc.; constant supervision	75
Fair control with weight batching	60
Poor control; volume batching of aggregates	40

by casting trial mixes, measuring their workabilities and cube crushing strengths, and then adjusting the mix accordingly. Much of this work can easily be performed in the laboratory. The part of the method with which *Table 2.2* is concerned has of course to be established by co-operation with the site. Considerable creditable research since *Road Note No. 4* has been faced with the inherent complexity of the problem and has not made this method obsolete as a useful simple method of designing a mix. All other methods can still be improved by studying trial mixes, as mentioned previously.

CP 114 used to specify concretes according to their minimum cube crushing strength at 7 and 28 days, and it is still possible for a designer to do this, but CP 110 has a more scientific approach—unfortunately more complicated. The mix design method presented in this book is based on the required average crushing strength. To design a mix with a certain specified minimum crushing strength, as for CP 114, we use *Table 2.2* to obtain the requisite average crushing strength, and then design a mix for this average crushing strength.

CP 110 specifies a concrete with a *characteristic strength*. For example, in Table 47 it defines Grades 20, 25, 30, etc., of concrete as having characteristic strengths of 20, 25, 30, etc., N/mm^2. If a large number of cubes of the same concrete (same age, curing, etc.) are tested the results can be plotted as shown in *Figure 2.4*. In statistics this figure is known as a *histogram*, and its shape is well known as a normal (gaussian) distribution. The average (or mean) cube strength is the value of cube strength corresponding to the centroid of this shape. As in this case we assume it to be a normal distribution, not skew[7] (though in fact it is slightly skew), the average (or mean) strength corresponds to the centre line of the shape, as shown. Statistical theory gives the formula:

Characteristic strength =

$$\text{Mean strength} - 1.64 \times \text{Standard deviation} \qquad (2.1)$$

The number 1.64 is derived from the fact that CP 110 chooses characteristic strength as the value below which we can expect 5 % of the cubes to fail (see *Figure 2.4*).

The breadth of the shape of *Figure 2.4* gives an indication of the scatter of the results. For statistical purposes this is expressed as *standard deviation*, which can be obtained thus: if we make n cube tests and their crushing strengths are $f_{cu1}, f_{cu2} \cdots f_{cun}$ then the mean crushing strength is $f_{cum} = (\Sigma f_{cu1})/n$ and the standard deviation is $\sqrt{[\Sigma(f_{cu1} - f_{cum})^2/(n-1)]}$.

If we are to design a concrete to a particular CP 110 characteristic strength then we must obtain the mean strength, i.e. Characteristic

c

strength + 1.64 × Standard deviation. Hence, we need to know the standard deviation. Equation 2.1 can be expressed as:

$$\text{Characteristic strength} = \text{Mean strength} - \text{Margin} \quad (2.2)$$

where

$$\text{Margin} = 1.64 \times \text{Standard deviation} \quad (2.3)$$

and thus

$$\frac{\text{Characteristic strength}}{\text{Mean strength}} = 1 - 1.64 \times (\text{Coefficient of variation})$$

$$(2.4)$$

where

$$\text{Coefficient of variation} = \frac{\text{Standard deviation}}{\text{Mean strength}} \quad (2.5)$$

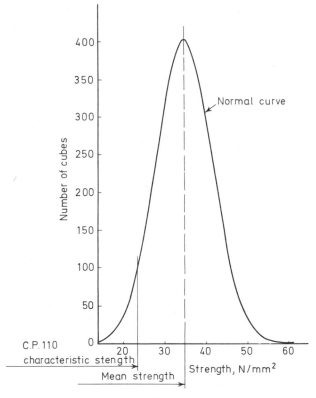

Fig. 2.4

For concretes stronger than 20 N/mm^2, CP 110 recommends (a) the standard deviation can be obtained for cube tests on at least 100 separate batches of concrete produced over a period of not more than one year, provided the margin is not less than 3.75 N/mm^2, or (b) the standard deviation can be obtained for cube tests on at least 40 separate batches of concrete produced over a period between 5 days and 6 months, providing the margin is not less than 7.5 N/mm^2, or (c) if histograms have not been established as for (a) and (b), then the margin can be simply taken as 15 N/mm^2.

2.3.10 *Design of concrete mix of given mean (or average) strength*

To design a concrete mix for industry the mean strength has first to be established as in Section 2.3.9. If however the mix is for a laboratory experiment then we design for the mean strength. The required water-to-cement ratio for the mean strength required is obtained from *Figure 2.2*, which assumes that the concrete is cured in air. Then a decision is made on the degree of workability, using *Table 2.1* as a guide. Then the most suitable aggregate-to-cement ratio can be chosen from *Table 2.3*. This table gives such ratios for different gradings (as given in *Figure 2.1*), workabilities, water-to-cement ratios, and types of aggregates.

Then if durability is important because, say, the concrete is exposed to injurious elements, i.e. is not protected inside an office block or laboratory, Tables 48 and 49 of CP 110 should be consulted to see if we have sufficient cement in our mix. If not, we decrease the aggregate-to-cement ratio accordingly. If this has to be done we might perhaps then repeat our design, taking advantage of say an increased workability to assist compaction and ease and therefore cost of concreting.

Example 2.1. To design a concrete mix for a pretensioned prestressed beam to have a mean (or average) crushing strength of 47 N/mm^2 at an age of 7 days.

The coarse aggregate to be used is a 19.05 mm ($\frac{3}{4}$ in) and down, rounded aggregate with a grading curve approximating to Curve 2 on *Figure 2.1*. Vibration is to be employed and the prestressing wires cause little obstruction to the placing of the concrete. We shall assume however that the beam is of I-section with narrow flanges and web. Hence it is decided that a *medium* workability is desirable (see *Table 2.1*).

Using rapid hardening Portland cement and consulting *Figure 2.2*, the necessary water-to-cement ratio is 0.35.

From *Table 2.3*, the aggregate-to-cement ratio is therefore 3.

To check that the cement content is adequate for durability in accordance

Table 2.3.

(1) 19.05 mm ($\frac{3}{4}$ in) rounded aggregate

Degrees of workability	Very low				Low				Medium				High			
Grading of aggregate*	1	2	3	4	1	2	3	4	1	2	3	4	1	2	3	4
Water-to-cement ratio by weight 0.35	4.5	4.5	3.5	3.2	3.8	3.6	3.2	3.1	3.1	3.0	2.8	2.7	2.8	2.8	2.6	2.5
0.40	6.6	6.3	5.3	4.5	5.3	5.1	4.5	4.1	4.2	4.2	3.9	3.7	3.6	3.7	3.5	3.3
0.45	8.0	7.7	6.7	5.8	6.9	6.6	5.9	5.1	5.3	5.3	5.0	4.5	4.6	4.8	4.5	4.1
0.50	—	—	8.0	7.0	8.2	8.0	7.0	6.0	6.3	6.3	5.9	5.4	5.5	5.7	5.3	4.8
0.55	—	—	—	8.1	—	—	8.2	6.9	7.3	7.3	7.4	6.4	6.3	6.5	6.1	5.5
0.60	—	—	—	—	—	—	—	7.7	—	—	8.0	7.2	×	7.2	6.8	6.1
0.65	—	—	—	—	—	—	—	8.5	—	—	—	7.8	×	7.7	7.4	6.6
0.70	—	—	—	—	—	—	—	—	—	—	—	—	×	—	7.9	7.2

(2) 19.05 mm ($\frac{3}{4}$ in) irregular gravel aggregate

Degrees of workability	Very low				Low				Medium				High			
Grading of aggregate*	1	2	3	4	1	2	3	4	1	2	3	4	1	2	3	4
Water-to-cement ratio by weight 0.35	3.7	3.7	3.5	3.0	3.0	3.0	3.0	2.7	2.6	2.6	2.7	2.4	2.4	2.5	2.5	2.2
0.40	4.8	4.7	4.7	4.0	3.9	3.9	3.8	3.5	3.3	3.4	3.5	3.2	3.1	3.2	3.2	2.9
0.45	6.0	5.8	5.7	5.0	4.8	4.8	4.6	4.3	4.0	4.1	4.2	3.9	×	3.9	3.9	3.5
0.50	7.2	6.8	6.5	5.9	5.5	5.5	5.4	5.0	4.6	4.8	4.8	4.5	×	4.4	4.4	4.1
0.55	8.3	7.8	7.3	6.7	6.2	6.2	6.0	5.7	×	5.4	5.4	5.1	×	4.8	4.9	4.7
0.60	9.4	8.6	8.0	7.4	6.8	6.9	6.7	6.2	×	6.0	6.0	5.6	×	×	5.4	5.2
0.65	—	—	—	8.0	7.4	7.5	7.3	6.8	×	×	6.4	6.1	×	×	5.8	5.6
0.70	—	—	—	—	8.0	8.0	7.7	7.4	×	×	6.8	6.6	×	×	6.2	6.1

(3) 19.05 mm (¾ in) crushed rock aggregate

Degrees of workability	Very low				Low				Medium				High			
Grading of aggregate*	1	2	3	4	1	2	3	4	1	2	3	4	1	2	3	4
Water-to-cement ratio by weight 0.35	3.2	3.0	2.9	2.7	2.7	2.7	2.5	2.4	2.4	2.4	2.3	2.2	2.2	2.3	2.1	2.1
0.40	4.5	4.2	3.7	3.5	3.5	3.5	3.2	3.0	3.1	3.1	2.9	2.7	2.9	2.9	2.8	2.6
0.45	5.5	5.0	4.6	4.3	4.3	4.2	3.9	3.7·	3.7	3.7	3.4	3.3	3.5	3.5	3.2	3.1
0.50	6.5	5.8	5.4	5.0	5.0	4.9	4.5	4.3	4.2	4.2	3.9	3.8	×	3.9	3.8	3.5
0.55	7.2	6.6	6.0	5.6	5.7	5.4	5.0	4.8	4.7	4.7	4.5	4.3	×	×	4.3	4.0
0.60	7.8	7.2	6.6	6.3	6.3	6.0	5.6	5.3	×	5.2	4.9	4.8	×	×	4.7	4.4
0.65	8.3	7.8	7.2	6.9	6.9	6.5	6.1	5.8	×	5.7	5.4	5.2	×	×	5.1	4.9
0.70	8.7	8.3	7.7	7.5	7.4	7.0	6.5	6.3	×	6.2	5.8	5.7	×	×	5.5	5.3

* Curve No. on *Figure 2.1.*
— Indicates that the mix was outside the range tested.
× Indicates that the mix would segregate.

with CP 110, Tables 48 and 49 give the minimum mass of cement per m³ of the concrete. Hence it is necessary to calculate this quantity from Section 2.3.12.

2.3.11 Combining aggregates to obtain a grading for the mix design method

Available sands and gravels need to be combined in suitable proportions so that the resultant grading approximates to one of the curves of *Figure 2.1*, so that the method of mix design of Section 2.3.10 can be used. A graphical method is given for obtaining these proportions in *Road Note No. 4*, but the method of calculation, illustrated by the following example, is simpler to explain and understand and the calculations are trivial.

Example 2.2. The gradings of sand and two coarse aggregates kept in stock in the concrete laboratory at Bradford University are given in Columns (a), (b) and (c) respectively of *Table 2.4*. Supposing these are combined to approximate to Curve 1 of *Figure 2.1*, whose grading is listed in Column (i) of *Table 2.4*.

To 1 kg of sand we can only decide how many kg x of 9.52 mm ($\frac{3}{8}$ in) gravel and how many kg y of 19.05 mm ($\frac{3}{4}$ in) gravel to mix with it to obtain the grading of Curve 1. Two unknowns only need two equations. Hence we can only make Curve 1 correct for the percentages passing two chosen sieve sizes. Suppose we choose the percentages passing apertures 9.52 mm ($\frac{3}{8}$ in) and 4.76 mm ($\frac{3}{16}$ in).

According to Curve 1, the percentage passing 9.52 mm ($\frac{3}{8}$ in) aperture is 45 %, hence using *Table 2.4*:

$$100 \times 1 + 96x + 19y = 45(1 + x + y)$$

According to Curve 1, the percentage passing 4.76 mm ($\frac{3}{16}$ in) aperture is 30 %, hence using *Table 2.4*:

$$100 \times 1 + 13x + y = 30(1 + x + y)$$

Solving these two equations, $x = 0.1172$, and $y = 2.345$. Thus the sand, 9.52 mm ($\frac{3}{8}$ in) gravel and 19.05 mm ($\frac{3}{4}$ in) gravel must be combined in the proportions 1:0.1172:2.345 respectively.

The grading of the combined aggregate is obtained by multiplying Columns (a), (b) and (c) of *Table 2.4* by 1, 0.1172 and 2.345 respectively, the products being shown in Columns (d), (e) and (f) respectively. The values in these columns are added together to give the values in Column (g) and then divided by $1 + 0.1172 + 2.345 = 3.462$ to give the values in Column (h), and this is the grading of the combined aggregate. Comparing this with Column (i) we have achieved the same percentages passing 9.52 mm and 4.76 mm apertures, as calculated. Our error is mainly for percentages passing 1.20 mm and 600 μm apertures. We could repeat the calculation say making the percentages passing apertures 9.52 mm and 1.20 mm equate in Columns

Table 2.4.

Aperture size	BS Sieve	Percentage passing (a) Sand	(b) 9.52 mm gravel	(c) 19.05 mm gravel	(d) (a) × 1	(e) (b) × 0.1172	(f) (c) × 2.345	(g) (d) + (e) + (f)	(h) (g) ÷ 3.462	(i) Curve 1
19.05 mm	$\frac{3}{4}$ in	100	100	97	100	11.72	227.5	339.2	98.0	100
9.52 mm	$\frac{3}{8}$ in	100	96	19	100	11.25	44.56	155.8	45.0	45
4.76 mm	$\frac{3}{16}$ in	100	13	1	100	1.524	2.345	103.9	30.0	30
2.40 mm	No. 7	85	1	0	85	0.1172	0	85.12	24.6	23
1.20 mm	No. 14	72	0	0	72	0	0	72	20.8	16
600 µm	No. 25	53	0	0	53	0	0	53	15.3	9
300 µm	No. 52	10	0	0	10	0	0	10	2.9	2
150 µm	No. 100	1	0	0	1	0	0	1	0.3	0

(h) and (i). Mix design is not a very accurate science and this is probably not worth the trouble and its result would not really be known to be any better. Various sets of two percentages passing certain sizes could be made equal in Columns (h) and (i) by calculation and all the various results plotted on a graph such as *Figure 2.1*, and one could choose the combined grading which looked generally closest to the graph of Curve 1. Again it is extremely doubtful if this is worth doing.

2.3.12 Quantities of materials required to make 1 m³ of concrete

A very simple method is illustrated in Example 2.3. This is useful for individual beams. If one needed considerable accuracy for a large quantity such as a dam, this can easily, and best, be established experimentally in the laboratory.

Example 2.3. Calculate the quantities of ingredients required for casting a beam and cubes in the laboratory having a total volume of 0.4 m³. The mix is to be in the proportions of 1 part cement to 0.87 parts sand to 0.10 parts 9.52 mm gravel to 2.03 parts 19.05 mm gravel by dry volumes with a water-to-cement ratio of 0.35 by masses.

Assume the bulk density of cement, sand and gravel to be 1440 kg/m³ (reasonably true if not using lightweight aggregates). Assume the density of the matured concrete to be 2400 kg/m³ (again reasonably true). The mass of the concrete is equal to the mass of its ingredients, except that much of the water will evaporate. Assume all the water vanishes—this will very slightly underestimate the cement, sand and gravel. Therefore:

1 kg cement	+ 0.87 kg sand	+ 0.10 kg small gravel	+ 2.03 kg large gravel	= 4 kg concrete

Mass of mature concrete = 0.4 × 2400 = 960 kg. Therefore requirements are 1 × 960/4 = 240 kg cement, 0.87 × 960/4 = 208.8 kg sand, 0.10 × 960/4 = 24 kg small aggregate, 2.03 × 960/4 = 487.2 kg large aggregate, 240 × 0.35 = 84 kg water. Then add 10% to these figures to allow for small underestimation and waste. This figure may need small adjustment according to experience of the concreting conditions, the particular mix and type of aggregates, etc.

2.3.13 Prescribed mixes

CP 110 gives prescribed mixes in Table 50 to replace the nominal mixes of CP 114. Generally these will give uneconomic concretes stronger than required. But they have the advantage that proper mix design procedures do not need to be established for the concreting plant.

2.3.14 Shrinkage

When cement, sand, gravel and water are mixed together the gross volume decreases as the finer particles arrange themselves in the interstices of the larger particles. This shrinkage continues as the concrete is being worked into place. Evaporation of water in the mix also decreases the volume of such concrete. It is possible to fill a mould, for example a 150 mm cube, and observe the concrete retract into the mould. Shrinkage, when the concrete is in a fluid state, does not matter structurally because no internal stresses can be instigated. There is an inaccurately known point at which the concrete changes from a fluid to a solid and immature fragile material. The exact time when this occurs depends upon the water-to-cement ratio, the type of cement, and the ambient humidity and temperature. After this time, further shrinkage of the concrete will cause internal stresses and even cracks to occur. The time when the transition occurs from liquid to solid is not precisely determinable, and it is difficult to know exactly when to commence measuring the shrinkage of the concrete in its solid state.

Measurements of the coefficient of shrinkage are possibly commenced too late to be of real mathematical value in research, because such readings are often commenced just when the specimen is hard enough to strip and handle for the purposes of the test. On such a

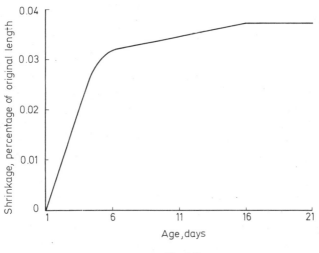

Fig. 2.5

basis the shrinkage coefficient can be of the order of 0.0005 at an age of 12 months and a typical relationship between shrinkage and age is illustrated in *Figure 2.5*. Initially the rate of shrinkage is high so that the error in not knowing the precise time to start measurements is quite appreciable. With the above coefficient, and supposing, for simplicity, the modulus of elasticity of concrete is $28\,000\,N/mm^2$, then if the concrete were restricted from shrinking the tensile stress induced in the concrete would be $0.0005 \times 28\,000 = 14\,N/mm^2$. The concrete would certainly crack as its ultimate tensile strength would only be about $2.8\,N/mm^2$. The coefficient of shrinkage is less for a lean mix than for a rich mix (in cement content). It is less for a low water-to-cement ratio than for a high one, is very sensitive to the method of curing, and is influenced to a lesser extent by all the other possible variables.

Shrinkage after the concrete has solidified continues as and when further water evaporates. The chemical reaction of cement with water, and thus the shrinkage, continues in the concrete seemingly indefinitely. A gel is formed which contracts upon desiccation and becomes very hard (see Section 2.1). If concrete is submerged in water this cement gel expands with considerable force, so that the whole mass of concrete expands. This expansion, however, can never equal the shrinkage which has already taken place. On drying the concrete in air, shrinkage again occurs. Therefore, when concrete is subjected to continual wetting and drying, as for example due to tidal action, it experiences corresponding expansions and contractions. If concrete is cast beneath water then it does not shrink at all but expands, owing to the cement gel absorbing water.

If a mass of concrete shrinks (or expands) uniformly and its movement is not restricted by any external forces, then no internal stresses can be induced in the concrete. This seldom happens in practice; usually any movement of the concrete is restricted internally by reinforcement embedded in the concrete, and often externally by its surroundings. Also, the surface of concrete will often dry out (and therefore shrink) faster than the internal particles of concrete. When the concrete of a reinforced beam is in the solid state, as it shrinks it also bonds to the reinforcement. The resistance of the reinforcement to contraction opposes the shrinkage of the concrete. Thus the concrete near to the reinforcement is in tension, a bond stress is developed between the two, and the reinforcement is in compression. Shrinkage cracks often exist in reinforced concrete beams at intervals along the length of the reinforcement. These are sometimes too small to be observed with the instruments normally available. When a reinforced concrete beam is tested, cracks can usually be observed at a lighter loading than predicted from the modulus of rupture of the concrete, indicating that cracks or tensile stresses are already present due to shrinkage.

Designs concerning conventional reinforced concrete work do not usually attempt to estimate the quantitative effect of shrinkage, because such calculations cannot be made with any degree of confidence and the basic assumptions of any mathematical analysis can be adversely criticised. Prestressed concrete designers simply treat shrinkage as a 'loss' reducing the prestressing force. The ultimate strength of a beam is not altered by shrinkage because when cracks occur the initial internal stress systems are released, yet shrinkage affects the size of cracks and deflections at working loads.

The particles at the surface usually experience different conditions of curing to internal particles. Their rates of shrinkage thus differ and this differential shrinkage can cause troublesome stresses, cracks and movements, e.g. the surface crazing of artificial stones and the warping of ground floor and road slabs. This effect can be reduced by endeavouring to cure the surfaces similarly to the internal fibres. The latter are fairly well sealed from the atmosphere so that to reduce differential shrinkage it is therefore desirable to seal the surfaces from the atmosphere. One way of achieving this is to immerse the concrete member in water for as long as possible. It is often more economical to cover with damp hessian sacks, sand, or waterproof sheets, or to spray periodically with water. A granolithic topping on a floor is very vulnerable to the detrimental effects of differential shrinkage and is usually kept damp for as long as practicable, and for at least seven days.

Shrinkage must always be borne in mind in the design and construction of structures. Whenever possible, concreting programmes aim at minimising the detrimental effects of shrinkage. For example, ground floor *slabs on solid* (placed over either suitable subsoil or suitably consolidated blinded hardcore), are often concreted in numerous independent portions each of about 4.5 m square, which are able to shrink before being joined together. Plain concrete roads are similarly constructed. This is not considered necessary when reinforcement is present. Numerous minute cracks are formed, but as the reinforcement resists shrinkage the overall contraction is negligible. Some engineers will attribute almost any serious crack in a structure solely to shrinkage. This is often a fallacy because the reinforcement of most structures has a considerable resistance to the forces exerted by the shrinkage of the concrete, so that shrinkage cracks in a long structure will take the form of very small cracks fairly regularly spaced throughout the length of the structure. A serious crack is more often caused by thermal expansion and contraction, and settlement.

2.3.15 Relationship between stress and strain for concrete

If a graph is plotted relating stress and strain, the shape of the curve obtained is very much influenced by the rate at which the stress is applied. It is also dependent upon the strength of the concrete under question and indeed to some degree upon all the other possible variables. *Figure 2.6* shows a relationship OAF which is typical of a

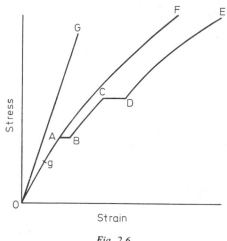

Fig. 2.6

concrete specimen loaded at a uniform rate. If the stressing had been held at the point A the concrete would have continued to strain under this particular constant stress. After a certain lapse of time, when the strain had reached the point B, had the stressing been recommenced at the previous rate, the relationship would have been the curve BC. Had the stressing been stopped at C, the same pheno-menon of *creep* would have occurred on C to D as occurred on A to B, that is the specimen strained or crept under constant stress until the stressing were recommenced at the point D, and the relationship then took the form represented by DE.

This phenomenon of creep (known in the U.S.A. as *plastic strain* or *time flow*) has been the subject of many investigations. *Figure 2.7* shows a curve CD which relates the creep (or strain) to time when the specimen is subjected to a constant stress. In this instance it took 5 s to apply the stress, so that the readings commenced from this time. It was once imagined that if this loading had been instantaneous and the observations of creep had been commenced immediately then this curve would have taken the form BCD. This is not so; the relationship is as ACD. Evans[8] constructed an apparatus which could load a specimen and record the strains at an extremely high speed. This enabled him to obtain readings of creep after an instantaneous loading to the stress in question, and enabled him to plot the curve AC in *Figure 2.7*. The same apparatus enabled him to discover an interesting relationship between stress and strain. At any particular stress an instantaneous increase in stress always gave a directly proportional increase in strain. Thus he obtained the linear relationship OG shown in *Figure 2.6*. This was an attempt to find a modulus

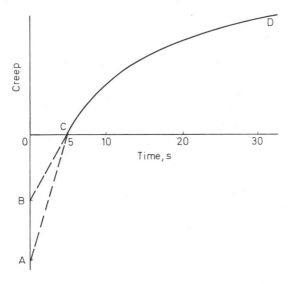

Fig. 2.7

of linear elasticity (Young's modulus) for concrete and thus to divorce the elastic from the plastic action, as in the early days attempts were made to use the elastic theories of design, which had been developed for steelwork, for reinforced concrete. This endeavour to separate elastic and plastic action was not subsequently favoured and creep cannot exactly be divorced from elasticity, shrinkage and other possible variables. Investigators generally agree that creep is mainly directly proportional to the constant stress causing it and proportional to a function of time. Various functions have been recommended for this.

It is thus distinctly noticeable that with regard to the relationship between stress and strain, concrete is comparable in behaviour to natural stones and timber, but certainly not to mild steel, because there is no period of proportionality, no marked elastic limit and no yield point. Apologies must therefore be made for using the term modulus of elasticity for concrete. However, from the early days this has been done in connection with the elastic theory which still has its uses. Therefore some value or values must be attributed to a rather mythical modulus of elasticity. *Figure 2.8* illustrates a typical stress–strain diagram for a concrete specimen and shows various ideas which have been propounded for the modulus of elasticity. OT_0 is tangential to the function at the origin and is called the *initial tangent modulus*. TPT′ is a tangent at the point P and is known as the *tangent modulus* at this point. Similarly $T_1QT_1′$ is the tangent modulus at point Q. The straight line PQ is called the *chord modulus* for the range P to Q. OP is the *secant modulus* for point P, and similarly OQ is the secant modulus for point Q. In *Figure 2.6* the slope of the curve OG is Evans' *short range* or *instantaneous modulus* of elasticity. This modulus is suitable for use in predicting the stresses caused in concrete structures by shocks from bombing or earthquakes.

The maximum permissible compressive stress in bending at working loads is often specified, for designs based on elastic theory, to be about one third of the crushing strength. Up to such working stresses the relationship between stress and strain approximates with reasonable accuracy to a straight line and most engineers utilise a secant modulus of elasticity corresponding to the maximum allowable working stress. This is the modulus of elasticity implied when reference is subsequently made to the modulus of elasticity of concrete, unless stated otherwise.

If points A and B in *Figure 2.6* were at the allowable working stress of the concrete under investigation, then the moduli of elasticity at points A and B are obviously different. One can take the modulus of elasticity for A and then make a separate calculation for creep. The former depends on the speed of loading to A, and the latter relies on debatable methods. It is usual, and simpler, to take the secant modulus of elasticity of point B, or whatever point on A to B one considers relevant to the time creep has been occurring. For example, concrete at the age of one year can have a modulus of elasticity of about one third of its value at the age of one month. When creep tests are made, specimens are cast out of the same mix, for the purpose of measuring the shrinkage which occurs. The shortening due to shrinkage can then be deducted, to give the true creep over the period independently of the effect of shrinkage.

Concrete made with certain popular lightweight aggregates can have a modulus of elasticity of only two thirds of the value of a conventional type of concrete of the same ultimate compressive

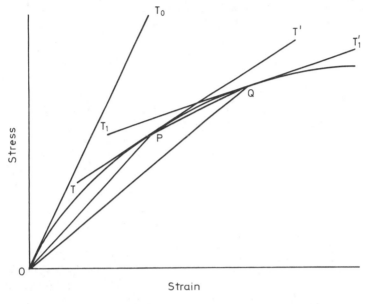

Fig. 2.8

strength. For elastic design the modulus of elasticity has been related to the concrete strength, but then for simplicity CP 114 adopted a constant value of 14 000 N/mm². For example, the modular ratio α_e used by CP 114, is Young's modulus for steel 210 000 N/mm² divided by 14 000, which equals 15.

If a reinforced concrete beam is subjected to a loading test α_e could be about 9 for use in calculations predicting deflection or stresses. If the load were maintained for say one year then α_e would be about 15.

With time, creep causes beams to deflect more, causes compression steel to be higher stressed, and causes long slender columns to increase their lateral deflection. This causes the bending moments to be higher and research by the author shows this to be a very important effect.

With regard to prestressed concrete, creep of steel (relaxation) and of concrete is calculated as a loss of prestressing force.

2.4 Types of reinforcement

Much reinforced concrete construction employs 'black' mild steel bars of circular cross section. In the early days, engineers often worried that such bars might not grip or bond to the concrete. Consequently, numerous bars were devised with surface deformations. As knowledge advanced, it became accepted that a mild steel bar of circular cross section could grip adequately to the concrete to develop its full tensile strength; surface deformations on the bar being superfluous.

Engineers generally are now happy to use high tensile steel provided the bar mechanically bonds with the concrete. If a mild steel bar of square cross section is twisted, this *cold working* converts it into a high tensile steel bar which can mechanically grip to the concrete. Another type of bar is made by rolling a round mild steel bar with a slight patterning on its surface, then subjecting it to cold working by tensioning and twisting to give a high tensile bar with a mechanical bond. Another type of high tensile bar is a hot rolled high tensile steel bar with a deformed surface. These high tensile reinforcements are called *high yield* by CP 110 because it is the yield stress which is of interest in our theories for ultimate strength. Cold working, for example, can increase the yield stress of mild steel much more than its ultimate stress. The advantage of using high yield bars is that the mass of steel required is reduced, and even though its cost per kilogram is higher than mild steel the total cost of the reinforcement and its fixing can be reduced. This does not apply in the case of the nominal reinforcement, which is usually more economic in mild steel, in a structure. Square twisted bars are bulkier for detailing, concreting, etc., than round deformed bars of the same strength. This disadvantage is reduced for a square twisted bar with chamfered corners. Sometimes square twisted bars have the advantage of bulk per unit cost for use as spacer bars in cylindrical shells—for keeping the fabrics apart and aiding concreting on the sloping surfaces. The appropriateness of a bar for a purpose and its cost and availability will usually decide which type of reinforcement to use. Mild steel is usually the most universally available and because more is required than high yield steel, say, as longitudinal tensile reinforcement in a cylindrical shell, then as Young's modulus is the same for both, the moment of inertia will be greater and hence the deflection less for shells with such mild steel. Also, the design has been elastic so that the lower strains of the mild steel do not conflict as much with the assumptions of the design. This also applies to frames which have their bending moments decided on elastic theory.

One should ascertain that any high yield reinforcement to be used bent does not have its strength seriously impaired by 'overstrain'. For example, the cold working of a bar introduces internal stresses in the bar. If the bar is then bent, further high stresses are superimposed on these stresses. It has been known for the fibres of steel on the inside of a bend to crush and for this not to be noticed until the bar was accidentally gently knocked, when the bar then came apart at the bend. Ref. 9 explains this problem and establishes that for two particular high yield bars, at that time, overstrain was not a practical worry. One of these bars had less cold working than the same make of bar at an earlier time. The amount of cold working is very important and a certain bar can have this altered for policy reasons from time to time without the designer necessarily realising that this has happened. A disadvantage of high yield bars is that the percentage of longitudinal tensile reinforcement is reduced, and it has been proved by many that this reduces the strength of a beam in shear. CP 2007 for water-retaining structures limits the stress in both mild and high yield steel to 84 N/mm^2 in the liquid-retaining face. The idea is that if the steel stress is kept low then the adjacent cracks in the concrete are similarly small. This is unfair to mechanically bonded bars, relative to plain bars, however, as research shows that at a given stress in the reinforcement the cracks will be more numerous and smaller for a mechanically bonded bar than for a plain bar. Certain recommendations for the design of structures to resist bombing do not allow high yield steels to be stressed as highly as mild steel reinforcements, because it is more brittle than mild steel, so that its strength can be impaired by sudden shocks.

High yield wires are used to make fabrics for reinforcing slabs (BS 1221). Cross wires are welded to the main wires and enable the main high yield wires to be mechanically bonded to the concrete. The chief advantage of such fabric reinforcements is the speed and low cost of fixing. A disadvantage is the high cost of fabrics. Also, fabrics do not commonly allow comparable economies to those effected by bending up or curtailing alternate bars in slabs. The steel over the supports of continuous slabs is far more rigid for concreting purposes when bars are used as opposed to fabrics. The main steel in a slab is sometimes inadequately anchored into the supporting beams when fabrics are used. The cross wires of BS fabrics do not satisfy the recommendation of CP 110, to the effect that the high yield reinforcement in any direction should be not less than 0.12 % of the gross cross-sectional area. Sometimes additional bars are laid on the fabric to supplement the area of the cross wires to comply with the recommendation, but quite often this has not been done. Such steel is important, however, when substantial temperature stresses are liable to occur or when the slab is of a substantial length (or width) in the direction of the cross wires.

In the U.K., wires commonly used for prestressed concrete are of 2, 5 and 7 mm diameter. Some are also available crimped or with indented surfaces. The wires usually need to be degreased before use

D

either with carbon tetrachloride or by allowing them to rust very slightly and then removing any loose rust. Some favour the latter with plain wires (ones not provided with a mechanical bond) so that the rust pitting can increase bond. The author consistently found both methods unsatisfactory for certain laboratory tests of beams with 2 mm diameter plain wires and reliably cured this trouble by using crimped wires. Strand is also very popular in the U.K.—this is essentially a wire rope. When stretched the wires tend to pull in laterally, resulting in a lower modulus of elasticity and also greater relaxation (creep) losses than with straight wires or bars. To reduce these disadvantages strand can be cold-drawn, which also makes it less bulky and stronger. Much work has also been done in the U.K. with high tensile steel bars having rolled-on threads. These threads do not weaken the bar like cut threads.

2.5 Practical use, creation and economics of structural concrete

Concrete is a heavy structural material. The largest spans of bridges are steel suspension bridges, next largest are steel trusses, steel girders, reinforced concrete arches, prestressed concrete girders, then reinforced concrete girders. Concrete is very cheap per unit compressive strength. This strength is weak relative to steel, so that in compression it has larger sections and does not have buckling problems as limiting as do steel columns and beams. This explains its economy for columns, arches and prestressed concrete, all essentially concrete in compression. Also many columns, say in a building, are within reason more economic than few, as the columns are more economic than longer span beams.

The large sections cause members to be heavy. It is important for economy to minimise the weight of suspended floors and roofs. Slabs cannot be too thin because of cracks due to shrinkage and temperature and thus the danger of a miscellaneous point load punching through. A minimum floor thickness is about 125 mm. For lightness and economy a floor 125 mm thick can be spanned continuously as far as possible and supported by T-beams which use the slab as their flanges. If the spans required are greater, then this system of beams can be supported by main T-beams. With this system, for economy, the length-to-breadth of the slab panels should be $\geqslant 2:1$. If less, then the slabs should be designed less economically to be two-way spanning. If because of supporting columns the grid of beams requires to be square, then two-way spanning slabs will be useful. If the overall floor thickness needs to be reduced then a flat-slab system may be used. Because of its shallow depth the amount of reinforcement needed is high and it is a heavy construction as none of the concrete not required in flexural tension is eliminated. Economy is improved in this latter respect by having dropped panels, but these can only be used economically for thicker floors of more than about 220 mm total thickness. Both types of flat slab have inexpensive shuttering but drop panels cause significantly more expense. As the self-weight is high they tend to be less economic for light loadings. Waffle floors can help the economy of this type of construction, but if the minimum crown thickness is too low and inadequately reinforced they can crack noticeably due to shrinkage, and for some structures this can interfere with serviceability.

Similar considerations apply to reinforced concrete roofs. The weight can be reduced by using shell roofs, and weight reduction is more important because the superimposed load is very light—even with a shell roof only 63 mm thick the self-weight is often 60% of the total load. The minimum thickness of a roof slab would be about 110 mm, and this plus supporting beams is far heavier than a shell roof.

Hollow tile roofs and floors are economic for *in situ* constructions where floors require to be say $\geqslant 200$ mm thick, and they can have the advantage of continuity and can provide flanges for T-beams.

The previous remarks apply to *in situ* concrete. Lightness and economy can be assisted by the use of precast concrete floor and roof units. Generally, they are less expensive than *in situ* floors and roofs, but the supporting beams lose efficiency and generally the structure is less robust. The great advantage and economy of continuity of beams and framing action of *in situ* work is reduced.

Prestressed concrete tends to be economic mainly when the depth allowed is inadequate for reinforced concrete construction.

The weight problem when overcome in a design automatically gives other advantages in the final structure, such as high natural frequency, easily spread small point loads and damping of small vibrations. Other advantages automatically obtained are good fire resistance and durability.

The structure is often dictated by client layout requirements. Aesthetics have not been mentioned because there are so many claddings and finishes available, e.g. a beam and slab floor often has a suspended ceiling to accommodate services so that the final appearance can be the same as a flat slab. Structural concrete usually looks best when the prime aesthetics of the building are based on the structure as opposed to the cladding. Both truism and proportioning according to strength requirements have parts to play, e.g. a pseudo-reinforced concrete shell roof composed of rolled steel girders and a

curved slab can look wrong and unattractive—the girders have
constant depth, looking too much in some places and too little in
others.

2.6 Bond between concrete and steel

This is a most necessary requirement of reinforced concrete
construction. If, for example, no bond existed between the tension
reinforcement of a beam and the surrounding concrete, then the
system would behave in the same way as a carriage spring, having two
leaves of different inertias and strengths, namely a relatively large
concrete leaf (possibly with a modulus of rupture of only say
3.5 N/mm^2) and a comparably small steel leaf (relatively strong with
a maximum ultimate fibre stress in bending of 520 N/mm^2). Under
these conditions the stiffer concrete member would resist most of the
superimposed bending moment and its ultimate strength would very
soon be realised, at such a load that the assistance of the reinforcement in resisting bending moment could be described as negligible.
Thus for the reinforcement to be utilised satisfactorily it has to bond
to the concrete so that a reinforced concrete beam bends as though
it is a homogeneous member (the strain in the reinforcement being
the same as the strain in the surrounding concrete fibres).

Pretensioned tendons must bond to the concrete which is cast
around them. Otherwise when released after the concrete has adequately matured, no precompression would be induced in the concrete,
the wires just sliding relative to the concrete.

Bond comprises two different actions. Firstly, there is the ability
of the concrete to stick to the steel. This is usually referred to as
adhesion. Secondly, there is the frictional resistance between the steel
and the concrete, often called *grip*. When a bar is tending to pull out
of its surrounding concrete the relative movement of the bar to such
concrete is known as *slip*. A bond stress cannot exist without its
coexistent strain, that is without slip. Adhesion is an initial resistance
to bond and occurs when the slip is minute. With a smooth cylindrical
bar, for example, adhesion is often attributed to micromechanical
locking (minute irregularities on the bar mechanically locking to the
concrete). As soon as a small amount of slip occurs the adhesion is
ruptured and takes no further part in the bond resistance. For such
slips a bond resistance is developed by the friction between the bar
and the surrounding concrete. This is aided by the shrinkage of the
concrete upon setting, as this causes the concrete to exert a radial
pressure on the reinforcing bar, thus increasing the frictional resistance between the two materials.

The frictional resistance can be assessed by multiplying such a pressure due to shrinkage by some suitable coefficient. Certain coefficients suggested by Armstrong[10] illustrate the sensitivity of the frictional resistance to the grease and rust on the surface of a bar. *Dilatancy* is a resistance to slip resulting from the wedging action of the small particles of concrete loosened after an initial slip has occurred. This effect constitutes a part of the general frictional resistance mentioned previously. The frictional resistance is enhanced at the locality of a crack where a *tangential friction* occurs because of the slight change in direction of the reinforcement bar.

Another contribution to the frictional resistance can be called *wedge action*. When the stress in a bar changes along its length due to its bond to the surrounding concrete, the effect of Poisson's ratio will cause a corresponding change in its cross-sectional area. Thus,

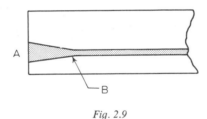

Fig. 2.9

such a reinforcement bar becomes slightly tapered and hence the term *wedge action*. With non-prestressed reinforced concrete this effect is extremely small. For prestressed concrete where steel stresses are much greater the wedge action is a significant asset. To illustrate this point, *Figure 2.9* exaggerates the effect; the pretensioned wire is unstressed after release at A and has therefore a larger diameter here than at B where the wire is in its fully stressed condition.

Both the adhesion and the frictional resistance are increased by *mechanical locking*, that is by using reinforcement bars with surface deformations which mechanically lock to the concrete[11].

Bond therefore consists of firstly an *adhesive resistance* and then a *frictional resistance*. As a simple illustration, *Figure 2.10* refers to a *pull-out test* of a steel rod from a concrete block. When the pull in the rod is P the portion of the graph AB represents the way in which the force in the rod is gradually transmitted to the concrete by frictional resistance. At the point B, the force still in the bar is insufficient to overcome the adhesive resistance of the remainder of the bar, and therefore BC represents the way in which the force in the rod is gradually transmitted to the concrete by adhesion. When the load in the bar is increased to P', the length of the bar slipping increases and the curve becomes A′B′C′, A′B′ being the frictional stage and B′C′ the adhesive stage.

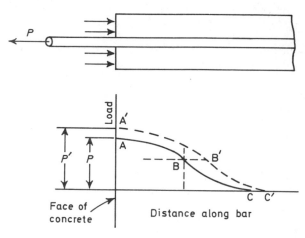

Fig. 2.10

When a reinforced concrete beam is subjected to bending, the tension reinforcement which is bonded to the concrete is such that both the steel and the concrete are in tension. This is a criticism of the above mentioned pull-out test in which the steel is in tension and its surrounding concrete is in compression. Tests[12] of bond stress are therefore made by measuring the strain in the steel, and the strain in the concrete touching such steel, along the lengths of the bars provided as tension reinforcement in beams.

2.6.1 Anchorage or bond length

Figure 2.11 shows a bar anchored into a block of concrete. The necessary bond length l_b is to be determined so that the bar can develop a tensile stress of f_s at section B. If the bar has a cross-sectional area A_s and perimeter u, then the force in the bar N_s is given by

$$N_s = A_s f_s \tag{2.6}$$

If f_{mbs} is the average bond stress between the steel and the concrete, this exists over an area of contact equal to ul_b, therefore

$$N_s = f_{mbs} u l_b . \tag{2.6a}$$

Eliminating N_s between these two equations

$$l_b = A_s f_s/(f_{mbs} u) \qquad (2.7)$$

If diameter of bar $= d_b$, then from equation 2.7

$$l_b/d_b = f_s/(4f_{mbs}) \qquad (2.8)$$

Table 2.5 enables anchorage lengths to be easily determined for bars in tension; values of the ratio l_b to d_b are read off for one's values of f_{cu} (= concrete grade or characteristic strength) and f_y (characteristic strength of steel). Values of ultimate anchorage bond stresses and f_y are from Table 22 of CP 110 and *Table 2.6* respectively. For example, for a plain bar and $f_{cu} = 25$ N/mm^2, $f_{mbs} = 1.4$, and if $f_s = f_y =$ say 250 N/mm^2 (mild steel) then from equation 2.8, $l_b/d_b = 250/(4 \times 1.4) = 44.6$, which is given as 45 in *Table 2.5*.

Table 2.5 Tension anchorage lengths (mm)

f_{cu}	20	25	30	$\geqslant 40$
f_y				
250	52	45	42	33
410	60	54	47	39
460	68	61	52	44
425	63	56	48	41
485	101	87	81	64
Stresses, N/mm^2		*Ratios of l_b to d_b*		

Table 2.6.

Designation	*Nominal sizes,* mm	f_y, N/mm^2
Plain hot rolled mild steel	all sizes	250
Deformed hot rolled high yield	all sizes	410
Deformed cold worked high yield	$\leqslant 16$	460
Deformed cold worked high yield	over 16	425
Plain hard drawn steel wire (fabrics)	$\leqslant 12$	485

When bars in compression are anchored, the compression on a bar is also resisted by the pressure on its end (e.g. end C in *Figure 2.11*). To allow for this it is simple to add a suitable amount to equation 2.6a, namely A_s times the compressive stress on the concrete. This was once done, but CP 110 (Table 22) prefers simply, but less logically and precisely, to increase the ultimate anchorage bond stresses for bars in compression. On this basis *Table 2.7* enables anchorage lengths to be easily determined for bars in compression, similarly to *Table 2.5* (see Section 2.6.5).

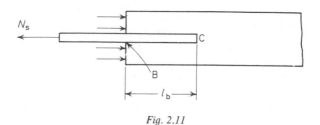

Fig. 2.11

Table 2.7 Compression anchorage lengths (mm)

f_{cu}	20	25	30	$\geqslant 40$
f_y				
250	42	37	33	27
410	49	43	38	32
460	55	48	43	36
425	51	44	39	33
485	81	71	64	53
Stresses, N/mm²	*Ratios of l_b to d_b*			

2.6.2 End anchorages

In practice reinforcement is seldom, if ever, perfectly clean of rust and/or mill scale and/or grease. This can have a more disastrous effect upon the anchorage in tension of plain than deformed bars. Hence it is good practice always to provide plain bars, when used in tension, with end anchorages such as hooks or nibs. These end anchorages are disadvantageous for deformed high yield steel because of cost, efficiency when stopping off bars in beams, and overstrain[9], but can be used if essential (e.g. lack of space in which to anchor at end of beam in some instances). Similarly it is disadvantageous in cost and efficiency to use end anchorages on bars in compression, but they can be used if essential. End anchorages are commonly hooks and nibs as shown in *Figures 2.12(a)* and *(b)* respectively. To anchor a bar, the overall length a required is the value of l_b from the tables of Section 2.6.1, less $16d_b$ and $8d_b$ for a mild steel hook and nib re-

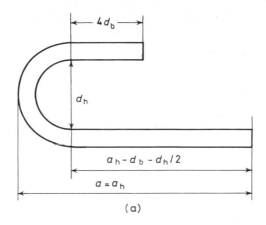

(a)

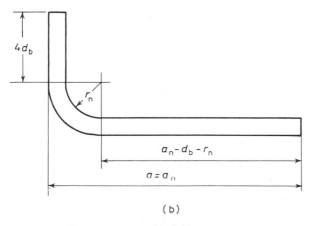

(b)

Fig. 2.12

spectively, and $24d_b$ and $12d_b$ for a high yield steel hook and nib respectively. After determining a for a bar we need to determine its total length. The total lengths of bars with hooks and nibs are $a_h + l_h$ and $a_n + l_n$ respectively. All these values are given in *Table 2.8* to aid designers. From the geometry of *Figure 2.12(a)*, the total length of the bar

$$= a_h + l_h = (a_h - d_b - 0.5d_h) + 0.5\,\pi\,(d_h + d_b) + 4d_b$$
$$\therefore\ l_h = 3d_b - 0.5d_h + 0.5\,\pi\,(d_h + d_b) \tag{2.9}$$

Table 2.8 Anchorage values of hooks and nibs

mm		d_b	6	8	10	12	16	20	25	32
Mild steel	*Hook*	$16d_b$	96	128	160	192	256	320	400	512
		$l_h (9d_b)$	54	72	90	108	144	180	225	288
	Nib	$8d_b$	48	64	80	96	128	160	200	256
		$l_n (5d_b)$	30	40	50	60	80	100	125	160
High yield	*Hook*	$24d_b$	144	192	240	288	384	480	600	768
		$l_h (11d_b)$	66	88	110	132	176	220	275	352
	Nib	$12d_b$	72	96	120	144	192	240	300	384
		$l_n (5.5d_b)$	33	44	55	66	88	110	138	176

From *Figure 2.12(b)*, the total length of the bar

$$= a_n + l_n = (a_n - d_b - r_n) + 0.5\,\pi\,(r_n + 0.5d_b) + 4d_b$$
$$\therefore l_n = 3d_b - r_n + 0.5\,\pi\,(r_n + 0.5d_b) \tag{2.10}$$

From these equations: for mild steel $d_h = 4d_b$ and $r_n = 2d_b$, thus $l_h = 8.85d_b$, say $9d_b$, and $l_n = 4.93d_b$, say $5d_b$; for high yield steel $d_h = 6d_b$ and $r_n = 3d_b$, thus $l_h = 11d_b$ and $l_n = 5.5d_b$.

Hooks are worth much more as an anchorage per unit length of material than nibs and cost little more to produce.

Tables 2.9 and *2.10* are based on *Table 2.6* and $f_{cu} = 20$ N/mm^2 and should be useful for designers of *in situ* concrete, because the weakest structural concrete is generally used for such work. If occasionally say $f_{cu} = 25$ N/mm^2 is used then if these tables are still used the bond lengths will be not unreasonably conservative. With regard to *Table 2.9*, plain mild steel bars are not recommended to be anchored without end anchorages and are therefore excluded from the table. The plain hard drawn fabric wires are, however, included as fabrics have welded cross wires which give extra security. Also see Section 2.6.5.

Example 2.4. A plain mild steel bar of 12 mm diameter is to be anchored with a hook. The characteristic strength of the concrete is 20 N/mm^2. Determine the overall length of the anchorage and the total length of the bar required for this anchorage.

From *Table 2.6*, $f_y = 250 \text{ N/mm}^2$.
From *Table 2.5*, $l_b/d_h = 52$, $\therefore l_b = 624$ mm.
From *Table 2.8*, $16d_b = 192$ mm, $l_h = 108$ mm
$$\therefore a_h = 624 - 192 = 432 \text{ mm}$$
$$\text{Total length} = 432 + 108 = 540 \text{ mm}$$

Alternatively, for these particular stresses, using *Table 2.10*, $a_h = 432$ mm, and from *Table 2.8*, $l_h = 108$, therefore total length $= 432 + 108 = 540$ mm.

Table 2.9 Straight anchorage lengths ($f_{cu} = 20 \text{ N/mm}^2$)

d_b, mm	6	8	10	12	16	20	25	32	f_y
$42\,d_b$	252	336	420	504	672	840	1050	1344	250
$49\,d_b$	294	392	490	588	784	980	1225	1568	410
$55\,d_b$	330	440	550	660	880	1100	1375	1760	460
$51\,d_b$	306	408	510	612	816	1020	1275	1632	425
$81\,d_b$	486	648	810	972	1296	1620	2025	2592	485
			Compression anchorage lengths (l_b), mm						N/mm^2
$60\,d_b$	360	480	600	720	960	1200	1500	1920	410
$68\,d_b$	408	544	680	816	1088	1360	1700	2176	460
$63\,d_b$	378	504	630	756	1008	1260	1575	2016	425
$101\,d_b$	606	808	1010	1212	1616	2020	2525	3232	485
			Tension anchorage lengths (l_b), mm						N/mm^2

Table 2.10 Overall anchorage lengths (mm) for hooks and nibs ($f_y = 250 \text{ N/mm}^2$, $f_c = 20 \text{ N/mm}^2$)

d_b	6	8	10	12	16	20	25	32
$a_h\,(36d_s)$	216	288	360	432	576	720	800	1152
$a_n\,(44d_s)$	264	352	440	528	704	880	1100	1408

2.6.3 Laps in reinforcement

To lap bars in compression, for example in columns, walls and sometimes over the supports of continuous T-beams, normally straight lengths are lapped the distance of the compression anchorage length (see Sections 2.6.1 and 2.6.5 and *Figure 2.13*). There is rarely

Table 2.11.

β	α_1						α_2						α_3						α_4					
	— Order of stopping-off or bending-up bars																							
	1st	2nd	3rd	4th	5th	6th	1st	2nd	3rd	4th	5th	6th	1st	2nd	3rd	4th	5th	6th	1st	2nd	3rd	4th	5th	6th
1	0	—	—	—	—	—	.11	—	—	—	—	—	0	—	—	—	—	—	.09	—	—	—	—	—
2	.15	0	—	—	—	—	.24	.11	—	—	—	—	.13	0	—	—	—	—	.21	.09	—	—	—	—
3	.21	.09	0	—	—	—	.30	.19	.11	—	—	—	.18	.08	0	—	—	—	.27	.16	.09	—	—	—
4	.25	.15	.07	0	—	—	.33	.24	.17	.11	—	—	.22	.13	.05	0	—	—	.30	.21	.15	.09	—	—
5	.28	.19	.12	.05	0	—	.35	.28	.21	.16	.11	—	.25	.15	.09	.04	0	—	.31	.24	.18	.13	.09	—
6	.30	.21	.15	.09	.04	0	.37	.30	.24	.19	.14	.11	.27	.18	.13	.08	.03	0	.33	.27	.21	.16	.12	.09
7	.31	.23	.17	.12	.08	.04	.39	.32	.26	.22	.17	.14	.29	.20	.15	.11	.07	.03	.34	.29	.23	.19	.15	.12
8	.32	.25	.19	.15	.10	.07	.40	.33	.27	.24	.20	.17	.30	.22	.17	.13	.09	.05	.35	.30	.25	.21	.18	.15

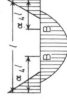

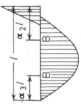

any advantage in using hooks or nibs and so reducing the lap length to the overall length of anchorage (see Section 2.6.2).

Lapping bars in tension is to be avoided. Plain bars without end anchorages should not be lapped in tension. When bars have to be lapped (see Sections 2.6.1, 2.6.2 and 2.6.5) in tension one should try to make laps, which need to be the distance of the tension anchorage length, as far from the places of maximum stress as possible and to stagger laps so that they do not overlap one another. For example, for a particular folded plate about 26 m long the tension steel to be used was in 12 m lengths. The number of bars of the same diameter which needed to be provided for the full length was increased by one; then each plain bar could be discontinued at any position. The system is indicated in *Figure 2.14*, bars A being of the maximum length possible and lengths a_h being the overall length of the hook anchorage. Adjacent hooks had a clear distance between them of about 75 mm to give a tolerance to the bar bender and fixer and to aid concreting.

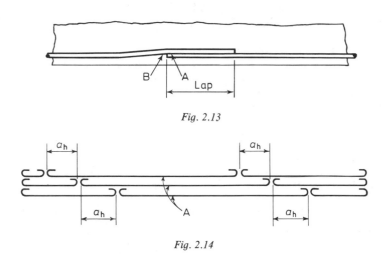

Fig. 2.13

Fig. 2.14

The compression lap shown in *Figure 2.13* should not be used in tension as the bars try and pull into line and thus outwards at A and B, trying to split off the concrete cover. If one is desperate to use this type of lap in tension, then the only chance of success is to use deformed bars and a stirrup at A, designed to resist the splitting force. The effective depth of reinforcement is reduced at B—to avoid this the lap shown can be rotated through a right-angle if detailing permits.

It is good practice to have a gap of about 15 mm between the lapped bars (*Figure 2.13*), to avoid voids in the concrete between the bars.

2.6.4 Curtailment of reinforcement in beams

Table 2.11 is useful for designers giving the points B where bars are
no longer required for resisting bending moment in a beam of span *l*.
It is based on uniformly distributed loads. For continuous spans it
assumes that the bending moments at mid span and support are
equal. Column β gives the number of bars at the position of maximum
sagging bending moment at or near to mid span. The coefficients α
are given for the order in which these bars are no longer required for
considerations of bending moment, counting from the position of
maximum sagging bending moment. The bending moment diagrams
to which the coefficients relate are shown below the table.

Strictly speaking, at the point when a bar is no longer required, if
it is not immediately bent-up for shear it can be just terminated, but
it must be checked that it has sufficient anchorage length to develop
its full tensile strength from the point where this is needed. However,
for plain bars a mechanical end anchorage is desirable (Section 2.6.2)
so the curve of the hook or nib can be commenced at this point where
the bar is no longer required.

Example 2.5. A simply supported beam carries a uniformly distributed load
over a span of 8 m and the design for ultimate limit state of bending requires
five 25 mm diameter deformed bars of hot rolled high yield steel in tension at
mid span. One of these bars is to be curtailed; determine the length of this
bar from mid span, assuming $f_{cu} = 20 \text{ N/mm}^2$.

From *Table 2.11*, $\alpha_1 = 0.27$, $\therefore \alpha_1 l = 0.27 \times 8 = 2.16$ m.
From *Table 2.6*, $f_y = 410 \text{ N/mm}^2$.
From *Table 2.9*, $l_b = 1500$ mm.

Allow no anchorage length but check that this bar has sufficient anchorage
length from mid span where it is fully stressed.

Length of bar from mid span $= 4 - 2.16 = 1.84$ m and this is all right,
as it is greater than 1.5 m. (Also see remainder of this section and Section 2.6.5.)

Example 2.6. In Example 2.5 now curtail a second bar. Determine its length
from mid span.

From *Table 2.11*, $\alpha_1 = 0.19$, $\alpha_1 l = 0.19 \times 8 = 1.52$ m.

From above $f_y = 410 \text{ N/mm}^2$, $l_b = 1.5$ m, and this bar is fully stressed
at $\alpha_1 l = 2.16$ m. Now $2.16 - 1.52 = 0.64$ m, which is less than 1.5 m and
thus inadequate anchorage. Length of this second bar from mid span is
thus $4 - 2.16 + 1.5 = 3.34$ m. (Also see remainder of this section and Section 2.6.5.)

But then CP 110 expresses concern that in practice the distribution
of live loading may not be as assumed and this would make errors
in the values of α. Hence it recommends an extra anchorage length be
added to each curtailed bar of $12d_b$ or its effective depth. Against this

is the fact that design loadings are sometimes very conservative and when the distribution is wrong the total is usually less.

CP110 also expresses concern about anchoring bars in tension zones and recommends bars extending 'an anchorage length appropriate to their design strength $(0.87\ f_y)$ from the point where they are no longer required to resist bending'. This seems very conservative relative to past practice and experience.

A method used successfully over many years by the author is simpler than the requirements of the preceding two paragraphs. It is based on the idea that the bar to be curtailed will be continued to some extent beyond the point where it is no longer required. There will thus be no sudden change in total tensile force on either side of this point, because the beam curvature and bending moment do not suddenly alter. Hence it is good practice to assume that all the bars have the same strain and stress at this point. Thus the bar to be curtailed is anchored for this stress, whether in a zone of tension or compression.

Example 2.7. If the 20 mm diameter bar is to be curtailed out of a group of two 25 mm diameter and one 20 mm diameter deformed bars, determine the length of this bar which must be continued past the point P where it is no longer required. Suppose for its design strength the 20 mm diameter bar needs an anchorage length of 1.05 m.

Tensile force required at point $P = 2 \times (\pi/4) \times 25^2 \times$ Design strength

$$\text{Stress in bars at this point} = \frac{2 \times (\pi/4) \times 25^2 \times \text{Design strength}}{2 \times (\pi/4) \times 25^2 + (\pi/4) \times 20^2} = 0.7576 \times \frac{\text{Design}}{\text{strength}}$$

Anchorage length required $= 0.7576 \times 1.05 = 0.796$ m.

Example 2.8. Repeat Example 2.5 with this alternative method.
As before $\alpha_1 l = 2.16$ m, $f_y = 410$ N/mm^2, $l_b = 1500$ m.
The anchorage length required from point $P = (4/5) \times 1500 = 1200$ mm.
Length of bar from mid span $= 4 - 2.16 + 1.2 = 3.04$ m, and this is all right as it is greater than 1.5 m.

2.6.5 Anchorage length reductions because of design strength being less than f_y

It has been assumed that a bar is anchored adequately to develop its full stress. This seems good practice. CP110 conservatively reduces the yield stress by a material factor, but then only requires anchorage for this reduced amount. This complicates matters, and reduces the anchorage lengths already given very slightly. If one wishes to take advantage of this, then:

1. *For tension reinforcement* in beams the design strength = $f_y/\gamma_m = f_y/1.15 = 0.87\, f_y$. Hence the anchorage lengths given may be reduced by 13%, or say 10% or $\frac{1}{8}$.

2. *For compression reinforcement* in beams the design strength = $f_y/(\gamma_m + 0.0005\, f_y) = f_y/(1.15 + 0.0005\, f_y)$. For simplicity the smallest denominator we are perhaps to use is, from *Table 2.6*, $1.15 + 0.0005 \times 250 = 1.275$. Hence it would be always within CP110 to take the design strength as $0.785\, f_y$. Hence the anchorage lengths given may be reduced by 21% or say 20% or 1/5.

It would be reasonable to ignore the refinements of this section, as this will still give much more economy than the very approximate methods suggested by CP 110 as alternatives to the full complexities described in this section.

2.6.6 Anchorage of bent-up shear bars

Bars bent up as shown in *Figure 3.5(b)* can be used as shear reinforcement. The anchorage length NBH is that required for the bar to be able to develop its design strength at the neutral axis N.

Example 2.9. A 25 mm diameter bar is bent up at 45° to resist shear. Its design strength is $f_y/\gamma_m = 250/1.15 = 217.4\ \text{N/mm}^2$, and $f_{cu} = 25\ \text{N/mm}^2$. The effective depth of the bottom tensile reinforcement = 450 mm, and the cover to the top steel = 25 mm. Determine the length BH.

In calculating BN we should use the depth of the neutral axis but for simplicity and slight extra safety we will use $0.5 \times 450 = 225$ mm. Then BN = $[225 - 25\,(\text{cover}) - 12\,(\text{half dia. bar})]\sqrt{2} = 266$ mm. The total anchorage length from *Table 2.5* is $45 \times 25 = 1125$ mm. We can either use this figure or economise further as in Section 2.6.5. If we do the latter (as for detailing such a shear reinforcement system, usually the shorter the length BH the better) the anchorage length becomes $1125/1.15 = 978$ mm. Economising further it has been common past practice to allow half the value of a nib for the anchorage effect of the bar deviating through 45° at B. CP 114 used to allow this, and it has some value, but it does not seem to be mentioned in CP 110. From *Table 2.8* this reduction in anchorage length is $200/2 = 100$ mm. From the same table the hook at H reduces the overall anchorage length by 400 mm. Hence $a_h = 978 - 100 - 400 = 478$ mm. Hence BH = $478 - \text{BN} = 478 - 266 = 212$ mm.

2.6.7 Bearing stresses inside bends

Figure 2.15 shows a reinforcement bar of diameter d in tension bent to any shape. At point P the tensile force in the bar is F_b and at point P' this force has become $F_b - \delta F_b$. This change δF_b is due to the bond stress over the length PP' shown as a force δF_b (this acts all around the

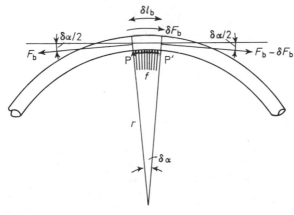

Fig. 2.15

perimeter of the bar). Because of the change of direction of the bar, and thus of the axial force in it, there is a bearing stress f inside the bend. Resolving forces perpendicular to PP'

$$f dr \delta\alpha = F_b \sin(\delta\alpha/2) + (F_b - \delta F_b)\sin(\delta\alpha/2)$$

In the limit when $\delta\alpha \to 0$, $\sin(\delta\alpha/2) \to d\alpha/2$, and $\delta F_b \to dF_b$.

$$\therefore 2 frd = 2F_b - dF_b$$

Now dF_b is negligible ($\to 0$) in comparison to the size of the quantities $2 frd$ and $2F_b$.

$$\therefore f = F_b/(rd) \tag{2.11}$$

This stress at P does not need to be checked for the standard anchorage hooks and nibs of Section 2.6.2. CP 110 requires f to be checked when the bar continues more than $4d$ after the bend and is still required for bond resistance—e.g. the bend at B in *Figure 3.5(b)* and at b' in *Figure 2.17*.

The same theory and equation (2.11) apply for bars in compression. In *Figure 2.15* F_b would be in the opposite direction and f would be at the opposite side of the bar.

A dowel bar under the bend just transmits and concentrates the

E

bearing stress to immediately below it, though it can help to spread this pressure transversely. CP 110 already does this considerably in its formula 3.11.6.8, and so such dowel bars are not considered helpful in reducing the bearing stresses inside bends.

Example 2.10. For Example 2.9 determine the minimum radius of curvature allowed at B. The beam is T-shaped and the bend B is in the wide flange.

Stress in bar at B = $217.4 \times (978 - 266)/978 = 158.3$ N/mm².

$d = 25$ mm, $F_b = 158.3 \times (\pi/4) \times 25^2 = 77690$ N.

The permissible f is, from CP 110 (formula 3.11.6.8), $a_b = \infty$,

$$= (1.5 \times 25)/(1 + 2 \times 25/\infty) = 37.5 \text{ N/mm}^2$$

Hence from equation 2.11

$$r = F_b/(fd) = 77690/(37.5 \times 25) = 82.87 \text{ mm}$$

2.6.8 Anchorage of stirrups (or links)

CP 110 recommendation 3.11.6.4 conflicts with the CP 110 recommendations already referred to in this chapter regarding anchorage length and bearing stress inside bends. Its inadequacy in this respect might be justified on the basis that the design of stirrups for shear is still conservative, but this is not indicated, and is not a sound approach. Against this, the anchorages are sometimes in tension zones and links are sometimes required to resist torsion, e.g. a beam of square cross section would experience maximum shear stress due to torsion not only at the neutral axis but at the centres of the top and bottom peripheries of the beam—where the links might have inadequate tension anchorage if in accordance with CP 110. Multitudes of beams in practice, designed with links to resist shear and not designed to resist torsion, do indeed have to resist varying amounts of torsion.

It would seem desirable[13] for the anchorages of links designed to resist shear and/or torsion to be in accordance with the previous sections of this chapter. In addition tests[9] show that deformed bars are only 10 % more effective in shear than plain bars, and that if the deformed bars are high yield then the failure is unexpected and violent. Deformed high yield stirrups should not be stressed any higher than mild steel links in shear[9]. This unfortunately disagrees with CP 110 but agrees with CP 114 (1957).

Example 2.11. Design the anchorage of an 8 mm diameter mild steel link of design strength $f_y/\gamma_m = 250/1.15 = 217.4$ N/mm², and $f_{cu} = 25$ N/mm². The internal dimensions of the link are 175×400 mm.

The tension lap from *Table 2.5* is $45d_b$. Two right-angle bends, using say the shape of link of *Figure 2.16*, are worth $8d_b$ each as anchorage (see Section 2.6.2). Hence tension lap required is $45d_b - 16d_b = 29d_b = 29 \times 8 = 232$

mm. The length of each vertical end is approximately (232 − 175) × 0.5 = 28.5 mm. This should be at least l_n of *Table 2.8*, i.e. $5d_b = 5 \times 8 = 40$ mm. The link is shown in *Figure 2.16*, the lap being along the top and down each side a length of 40 mm.

2.6.9 Splitting effects of bar anchorages

Anchoring a bar abcd from a beam into a column as shown in *Figure 2.17* is bad practice, causing splitting of the column along bcd. Even if the bearing stress is in order at b, increasing the length bcd

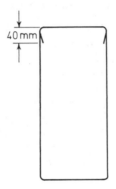

Fig. 2.16

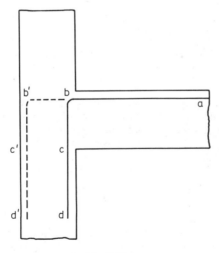

Fig. 2.17

does not add useful anchorage length, because of the splitting weakness. The bar should be taken as far across the column as possible, i.e. ab'c'd'. Designs are made for bending moments and shear forces assuming members to be concentrated at their centre lines. The true internal stress system at a practical junction is difficult to assess, hence the junction should be detailed as conservatively as possible, i.e. bb' should be as great as possible. In calculating the anchorage length, the bend at b' is worth the values $8d_b$ and $12d_b$ of nibs in *Table 2.8*.

2.6.10 Anchorage lengths based on elastic analysis

Equation 2.8 can be used provided f_s is taken as the permissible stress for a bar in tension and f_{mbs} as the permissible bond stress. Permissible stresses are stresses at working loads and are given in CP 114 and CP 2007. In a similar way Sections 2.6.2–2.6.10 (excluding Section 2.6.5) apply.

Thus CP 2007, Part 2, 1970, gives $f_s = 85$ N/mm² in tension and $f_{mbs} = 0.90$ N/mm² and 0.83 N/mm² for plain bars and 1:1.6:3.2 concrete and 1:2:4 concrete respectively. The respective anchorage lengths are thus $85d_b/(4 \times 0.9) = 23.61d_b$ and $85d_b/(4 \times 0.83) = 25.6d_b$. *Table 2.12* is to help designers of water containers.

Table 2.12.

d_b	mm 6	mm 8	mm 10	mm 12	mm 16	mm 20	mm 25	mm 32
23.6 d_b	142	189	236	283	378	472	590	756 mm
25.6 d_b	154	205	256	307	410	512	640	819 mm

Example 2.12. Determine the overall anchorage length of a 20 mm diameter plain bar of mild steel with an end hook, permissible tensile stress = 85 N/mm² and permissible average bond stress = 0.90 N/mm².

From *Table 2.12*, straight anchorage length = 472 mm.

From *Table 2.8*, hook is worth 320 mm.

Hence (see *Figure 2.12(a)*) $a_h = 472 - 320 = 152$ mm.

REFERENCES

1. Bogue, R. H., *Chemistry of Portland Cement*, Reinhold, New York (1955)
2. Troxell, G. E., Davis, H. E., and Kelly, J. W., *Composition and Properties of Concrete*, McGraw-Hill, U.S.A. (1968)

3. Brunauer, S., 'Tobermorite Gel—The Heart of Concrete', *Am. Scientist*, Mar. (1962)
4. Neville, A. M., *Properties of Concrete*, Pitman (1963 and 1973)
5. Robson, T. D., 'High Alumina Cements and Concretes', *Contractors' Record* (1962)
6. Bate, E. E. H., and Stewart, D. A., 'A Survey of Modern Concrete Technique', *Proc. I.C.E.* Part 3, Dec. (1955)
7. Hughes, B. P., *Limit State Theory for Reinforced Concrete*, Pitman (1971)
8. Evans, R. H., 'Effect of Rate of Loading on Some Mechanical Properties of Concrete', Proceedings of the 1958 London Conference organised by the Mining Research Establishment of the N.C.B. in consultation with the Building Research Establishment, D.S.I.R., pp. 157–175
9. Wilby, C. B., 'Overstrain in High-Tensile Reinforcing Bars at Bends and in Stirrups', *Indian Concrete Journal*, Jan. (1962)
10. Armstrong, W. E. I., 'Bond in Prestressed Concrete', *Journ. Inst. Civil Engrs.*, 33, Nov. (1949)
11. Regan, P. E., and Yu, C. W., *Limit State Design of Structural Concrete*, Chatto and Windus (1973)
12. Evans, R. H., and Robinson, G. W., 'Bond Stresses in Prestressed Concrete from X-ray Photographs', Paper No. 6025, *Proc. I.C.E.*, Part 1, Mar. (1953)
13. Evans, R. H., and Wilby, C. B., *Concrete: Plain, Reinforced, Prestressed and Shell*, Art. 5.7, Ed. Arnold (1963)

Chapter 3

Reinforced concrete beams

3.1 Design

To design a reinforced concrete beam a reasonable procedure is as follows:

1. Estimate the dimensions of the beam. The overall depth can be taken as say a proportion of the effective span, 1/20 for simply supported, 1/25 for continuous, and 1/10 for cantilever beams. The breadth (or breadth of rib) can be taken as $\frac{1}{3}$ to $\frac{1}{2}$ of this depth.

2. If a rectangular beam, the ratio of the maximum distance between lateral restraints to breadth is ideal if less than 30, reasonable if between 30 and 40, likely to be impracticable if more than 50. This is because of the possibility of narrow beams buckling sideways.

3. Check the strength in shear, and torsion if present, in the worst case, usually a section adjacent to a support. It may well be that reinforcement is required and this should not normally be greater than say 10 mm diameter stirrups (two, four or six arm according to width of beam) at 80 mm centres. The beam may well eventually be detailed with bent-up bars assisting the stirrups in the localities of maximum shear force.

4. Check the strength in bending. For a rectangular (or simply supported) beam this is done at the section subjected to maximum bending moment. For a continuous T- or L-beam, mid span will normally be all right if the supports are, because the beam is acting as a rectangular beam at the supports. If compression steel is required it might be desirable because of detailing to revise the design to eliminate the need for such steel—if this is not done it is useful practically for it not to exceed 1 % of the breadth (of rib for a T-beam or L-beam) times the overall depth. Then determine the longitudinal tension reinforcement and see if it can be detailed reasonably in the beam. In the case of simply supported T- and L-beams it is speedier to calculate the reinforcement before the compressive strength of the section.

The beam has now been reasonably well designed and it is now only a matter of checking the limit states of deflection and cracking.

3.2 Elastic analysis for bending moments

At working loads the elastic analysis gives a reasonably accurate assessment of the (longitudinal) stresses in the concrete and reinforcement. It also gives a reasonable assessment of deflection experienced in a loading test using a Young's modulus obtained from specimens of the concrete. For the deflection of a member in practice, for say $1:2:4$ or $1:1\frac{1}{2}:3$ concrete, the elastic analysis is reasonable for design purposes if a Young's modulus of the concrete of say $14 \, \text{kN/mm}^2$ is used for continuously sustained loading and $21 \, \text{kN/mm}^2$ is used for loading of short duration.

CP 114 allowed the design of beams to be based on elastic analysis, restricting stresses to well within the elastic behaviour of the materials at working loads. Multitudes of structures which have lasted many years illustrate the safety of such designs. This method has been dispensed with by CP 110, but is still used for water-retaining structures (CP 2007).

For shell roofs the analysis for forces and bending moments is elastic, so it would seem logical and safe (because our experience is based on elastic design) to use elastic analysis for designing for these forces and moments. Where experimental evidence has not adequately ratified the methods of predicting ultimate bending moments, members can be designed by elastic theory with confidence, e.g. shells, a beam with unsymmetrical section with skew loading, etc.

For the above reasons the elastic analysis will be presented as concisely as possible, using the moment of inertia of the equivalent concrete section method only.

3.2.1 Assumptions made in the elastic design of reinforced concrete

Firstly, it is assumed that *plane sections subjected to bending remain plane after bending* (Bernoulli's theorem). This is found to be reasonably true by experiment, and means that the *distribution of strain is linear*.

It is also assumed that *stress is proportional to strain for both the steel and the concrete*. This is accurately true for the steel up to the limit of proportionality, but only approximately true for the concrete as far as the allowable working stress, and is most inaccurate above this stress towards failure. The elastic method of design endeavours to compute the stresses at working loads, and limits these stresses to amounts dependent upon the yield stress of the steel and the crushing stress of the concrete. The respective factors of safety are obtained from experience in industry. It can therefore be appreciated that beams designed in such a fashion are safe but have varying load factors on their ultimate strengths. CP 110 prefers to specify a load factor on the ultimate strength of a beam. Advocates of elastic design feel that stresses and therefore the size of cracks at working loads are controlled.

Perfect bond is assumed between the steel and the concrete. The concrete shrinks upon setting and therefore exerts a pressure upon the steel, which assists the resistance to friction between the two materials. This pressure is reduced to some extent when the steel and surrounding concrete are stressed in tension because Poisson's ratio is greater for steel (approximately 0.29) than for concrete (approximately between 0.20 and 0.14). The converse applies when the steel and surrounding concrete are stressed in compression. Irregularities on the surface of the reinforcement lock the steel mechanically to the concrete. Several proprietary high tensile bars and prestressing wires are purposely manufactured to create such an effect.

The *depth of the steel reinforcement is considered to be negligible compared with the depth of the beam.* This is usually a reasonable assumption.

Normally, temperature and shrinkage stresses are ignored in the design of sections to withstand bending moments, shear forces, and axial forces. It can be mentioned here that fortunately the thermal coefficients of expansion of concrete and steel are sensibly the same. For the design of the structure as a whole, temperature and shrinkage effects must be considered. For example, temperature stresses are particularly important in the design of chimneys, and losses in prestress due to shrinkage are important.

Concrete is assumed to be cracked in tension when bending stresses are considered. This is because the tensile strength of concrete is only about one tenth (and can be as little as one thirtieth for high strength concretes) of its compressive strength. The same concrete is, however, expected to resist diagonal tensile stresses. If the beam were prestressed it would be permissible for certain small tensile stresses to occur under bending. CP 2007 for the design of water-retaining structures assumes that the concrete will withstand tensile stresses so that no cracks occur, but nevertheless does not trust the concrete in tension structurally. In fact concrete has a most unreliable resistance to tension. The ultimate strengths of numerous direct tension specimens made from the same batch of concrete in an exactly similar fashion can vary enormously. The maximum strength can often be as much as twice the minimum strength. The ultimate tensile stress in bending, judged on the extreme fibre stress, using the assumptions of the elastic analysis, (and known as the *modulus of rupture*) is higher and more reliable than the direct tensile strength.

Concerning the design of prestressed concrete beams in bending the modulus of elasticity for concrete in tension is assumed to be the same as the value of this modulus for concrete in compression.

3.2.2 *Moment of inertia of reinforced concrete section*

Referring to *Figure 3.1(a)*, XX is the neutral axis of any section subjected to bending, δA_{c1} is a small portion of area of the concrete at a distance d_{c1} from the neutral axis, and δA_{s1} is a small portion of area of the steel at a distance d_1 from the neutral axis.

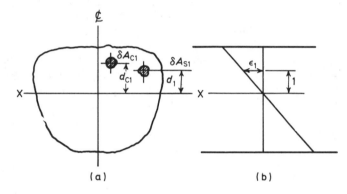

Fig. 3.1

The distribution of strain is linear and is shown in *Figure 3.1(b)*. Let the strain be of magnitude ε_1 at unit distance from the neutral axis. Therefore

$$\text{Strain for portion } \delta A_{c1} = \varepsilon_1 d_{c1}$$
$$\therefore \text{ Stress for portion } \delta A_{c1} = \varepsilon_1 d_{c1} E_c$$

If E_c and E_s are the Young's moduli for the concrete and steel respectively, the force for portion $\delta A_{c1} = \varepsilon_1 d_{c1} E_c \delta A_{c1}$ and similarly the force for portion $\delta A_{s1} = \varepsilon_1 d_1 E_s \delta A_{s1}$. Therefore, the moment of resistance of the section, M, is

$$M = \Sigma(\varepsilon_1 d_{c1} E_c \delta A_{c1})d_{c1} + \Sigma(\varepsilon_1 d_1 E_s \delta A_{s1})d_1$$
$$\therefore M = \varepsilon_1 E_c(\Sigma \delta A_{c1} d_{c1}^2 + \Sigma \alpha_e \delta A_{s1} d_1^2) \tag{3.1}$$

where $\alpha_e = E_s/E_c$ is the *modular ratio*.

Comparing equation 3.1 with the classical formula $M = f(I/y)$, where f is the stress at distance y from the neutral axis, $f/y = \varepsilon_1 E_c y/y = \varepsilon_1 E_c$ if we consider concrete only, in which case I is the *equivalent moment of inertia* (or *second moment of area*) of the cross section. Hence

$$M = \varepsilon_1 E_c I \tag{3.2}$$

and comparing this with equation 3.1

$$I = \Sigma \delta A_{c1} d_{c1}^2 + \Sigma \alpha_e \delta A_{s1} d_1^2 \tag{3.3}$$

The area of steel δA_{s1} can be regarded as equivalent to an area of concrete $\alpha_e . \delta A_{s1}$. In other words $\alpha_e . \delta A_{s1}$ is the *equivalent area* of the area of reinforcement δA_{s1}. This means that to obtain I we just multiply each steel area by α_e and then obtain the moment of inertia of the section as though it were all of concrete. If is often convenient when considering compression steel to consider the gross section of concrete and, as the area of the compression steel has not been subtracted, to multiply each of the steel areas by $(\alpha_e - 1)$ instead of α_e. These give the areas, in excess of the gross area, due to steel.

Example 3.1. The section shown in *Figure 3.2(a)* resists a bending moment of 56 kN m. Determine the maximum stress in the concrete and the stress in the steel if $\alpha_e = 15$.

Equivalent area of steel $= 15 \times 2 \times 0.7854 \times 25^2 = 14\,730$ mm^2. *Figure 3.2(b)* shows equivalent area of section, and centroid of this is the neutral axis XX. Equating moments of equivalent areas about axis XX

$$(150x)(x/2) = 14\,730\,(450 - x)$$

$$\therefore x^2 + 196.4x - 88\,380 = 0$$

$$\therefore x = 214.9 \text{ mm}$$

Taking moments of (equivalent) area about XX

$$I = (150x^3/3) + 14\,730 \times (450 - x)^2$$

$$= 1310 \times 10^6 \text{ mm}^4$$

From equation 3.2 $\varepsilon_1 = M/(IE_c) = (56/1310)E_c = 0.042\,75/E_c$. *Figure 3.2(c)* gives distribution of strain and *Figure 3.2(d)* gives corresponding distribution of stress. Therefore

$$f_c = E_c(\varepsilon_1 x) = 0.042\,75 \times 214.9 = 9.187 \text{ N/mm}^2$$

and

$$f_s = E_s \varepsilon_1 (450 - x) = 0.042\,75\,\alpha_e (450 - x) = 150.8 \text{ N/mm}^2$$

These last two equations are sometimes expressed as

$$f_c = Mx/I \text{ and } f_s = \alpha_e (M/I)(450 - x) \tag{3.4}$$

Example 3.2. If the beam of Example 3.1 were simply supported over an (effective) span (*l*) of 9.75 m and all the loading was uniformly distributed (*q*), determine the central deflection. Assume that the bending moment of 56 kN m was at mid span. Take $E_s = 200 \, kN/mm^2$; then $E_c = 200/\alpha_e = 13.33 \, kN/mm^2$.

Central deflection at mid span = $a_1 = (5/384)(ql^4/EI)$
In this example $M = ql^2/8 \therefore q = (56 \times 8)/9.75^2 = 4.713 \, kN/m$ (or N/mm)

$$\therefore a_1 = \frac{5 \times 4.713 \times 9750^4}{384 \times 13\,330 \times 1310 \times 10^6} = 31.76 \, mm$$

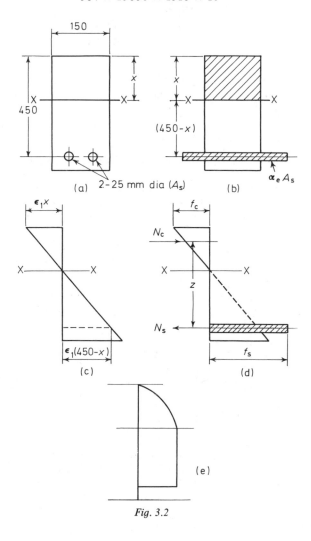

Fig. 3.2

Example 3.3. Determine the moment of resistance of the section shown in *Figure 3.2(a)* if the permissible stresses (i.e. the stresses allowed at working loads) are 10.5 N/mm^2 and 210 N/mm^2 for the concrete and steel respectively, and the modular ratio is 15.

From Example 3.1, $x = 214.9$ mm and $I = 1310 \times 10^6$ mm^4.

If concrete is the criterion, from equation 3.4

$$\text{Moment of resistance} = f_c(I/x) = 10.5 \times 1310 \times 10^6/214.9 \text{ N mm}$$

$$= 64.01 \text{ kN m}$$

If steel is the criterion, from equation 3.4

$$\text{Moment of resistance} = \frac{f_s}{\alpha_e} \cdot \frac{I}{(450 - x)} = \frac{210 \times 1310 \times 10^6}{15(450 - 214.9)} \text{ N mm}$$

$$= 78.01 \text{ kN m}$$

Therefore according to the assumptions of this design the moment of resistance of the beam is limited by the compressive strength of the concrete to 64.01 kN m.

3.2.3 Method for tabulating calculations for x and I

Table 3.1 illustrates the method. A is the equivalent area of a portion, y is the distance of the centroid of A from any chosen axis, say YY for *Figure 3.3(a)*, I_n is the second moment of area for the portion about its

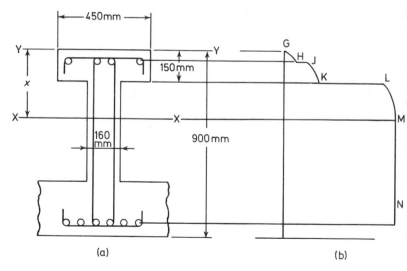

(a)   (b)

Fig. 3.3

Table 3.1.

Portion	Area	A	y	Ay	Ay²	I_n
Concrete	(45 − 16) × 15 = 435	435	7.5	3263	24 470	$29 \times 15^3/12 = 8156$
Concrete	16x	16x	0.5x	$8x^2$	$4x^3$	$16x^3/12 = 1.333x^3$
Compression steel	4 × 3.142 = 12.57	(×14 =) 175.9	5	879.6	4398	—
Tensile steel	6 × 4.909 = 29.45	(×15 =) 441.8	84.7	37 420	3 170 000	—
Totals		16x+ 1053		$8x^2 +$ 41 560	$5.333x^2 +$ 3 207 000	

neutral axis. Then taking moments of area about YY

$$\Sigma Ay = x\Sigma A \qquad (3.5)$$

$$\therefore x = \Sigma Ay/\Sigma A \qquad (3.6)$$

Second moment of area of whole section about YY

$$= I_y = \Sigma Ay^2 + \Sigma I_n \qquad (3.7)$$

If I is second moment of area of the whole section about its neutral axis XX, and x is depth of neutral axis below YY, then

$$I_y = x^2\Sigma A + I \qquad (3.8)$$

From equations 3.7 and 3.8

$$I = \Sigma Ay^2 + \Sigma I_n - x^2\Sigma A \qquad (3.9)$$

Supposing we wish to obtain the lever arm z. Then considering the tensile steel, area A_s, effective depth d_1, the moment of resistance $= f_s A_s z = f_s I/(d_1 - x)$

$$\therefore z = I/[A_s(d_1 - x)] \qquad (3.10)$$

Example 3.4. The section shown in *Figure 3.3* is through the counterfort of a tank. The reinforcement bars are of 25 mm diameter in tension and 20 mm diameter in compression and have 40 mm cover of concrete, $\alpha_e = 15$, the permissible stresses are: concrete in compression 8.3 N/mm^2, and steel in tension 85 N/mm^2. Determine the moment of resistance of the section at working loads, and the stress in the compression steel.

Table 3.1 shows the calculation (dimensions are in centimetres for convenience). Then from equation 3.5

$$8x^2 + 41\,560 = 16x^2 + 1053x$$

$$8x^2 + 1053x - 41\,560 = 0$$

$$\therefore x = 31.79 \text{ cm} = 317.9 \text{ mm}$$

From equation 3.9

$$I = 5.333x^2 + 3\,207\,000 - x^2(16x + 1053)$$

$$= 3\,207\,000 - 31.79^2 (16 \times 31.79 + 1048)$$

$$= 1.634 \times 10^6 \text{ cm}^4 = 16\,340 \times 10^6 \text{ mm}^4$$

Referring to equations 3.4:

Moment of resistance (concrete) $= 8.3 \times \dfrac{16\,340 \times 10^6}{317.9}$ N mm $= 426.6 \text{ kNm}$

Moment of resistance (steel) $= \dfrac{85}{15} \times \dfrac{16\,340 \times 10^6}{(847 - 317.9)}$ N mm $= 175.1 \text{ kN m}$

The latter is therefore the criterion, and the stress in the compression steel will be

$$= 15 \times \frac{175.1}{16\,340} \times (317.9 - 50) = 43.05 \text{ N/mm}^2$$

The latter is well within the permissible given by CP 2007, and 426.6 is much greater than 175.1, hence a designer might reduce the diameter of the 20 mm bars—unless their robustness is required to support the reinforcement cage. Their number cannot be reduced because of the stirruping system.

3.2.4 Popular formulae for slabs and rectangular beams (*elastic theory*)

For a rectangular section such as shown in *Figure 3.2*, if b is its breadth and d the effective depth of the tension steel, then moments of areas about XX give

$$bx^2/2 = \alpha_e A_s (d - x)$$

Dividing throughout by bd^2 and substituting $\rho = A_s/bd$ and $x_1 = x/d$

$$x_1^2/2 = \alpha_e \rho (1 - x_1) \tag{3.11}$$

$$x_1^2 + 2\alpha_e \rho x_1 - 2\alpha_e \rho = 0$$

$$x_1 = -\alpha_e \rho + \sqrt{[(\alpha_e \rho)^2 + 2(\alpha_e \rho)]} \tag{3.12}$$

As strain is linear, from *Figure 3.2*

$$\varepsilon_1 = f_c/(E_c x) = f_s/[E_s(d - x)] \tag{3.13}$$

$$\therefore f_s/f_c = \alpha_e (d - x)/x = \alpha_e (1 - x_1)/x_1$$

Let $\alpha_f = f_s/f_c$, then

$$x_1 = \alpha_e/(\alpha_e + \alpha_f) \tag{3.14}$$

From Equations 3.11 and 3.13

$$\rho = x_1/2\alpha_f \tag{3.15}$$

In *Figure 3.2(d)* N_c is the total force ($= 0.5 f_c bx$) of the compressive stress in the concrete, and N_s is the force ($= A_s f_s$) in the tension steel. The distance between these two forces z is called the *lever arm* or *moment arm*, and $z_1 = z/d$. Thus

$$z = d - x/3 \text{ or } z_1 = 1 - x_1/3 \tag{3.16}$$

Resolving longitudinally $N_c = N_s$. If M is the bending moment resisted by the section then $M = N_c z = N_s z$, thus

$$M = N_c z = 0.5 f_c bxz = (0.5 f_c x_1 z_1) bd^2 = Kbd^2 \tag{3.17}$$

where $K = 0.5 f_c x_1 z_1 = M/bd^2$. Also

$$M = N_s z = A_s f_s z_1 d \tag{3.18}$$

Designers make use of the full permissible stresses of concrete and steel (unless other factors (e.g. deflection) dictate otherwise), and then the previous equations give useful design formulae. For example, for water containers from CP 2007 the permissible stresses in concrete (1:1.6:3.2 mix) and steel are 8.3 N/mm^2 and 85 N/mm^2 respectively, and $\alpha_e = 15$. Substituting these figures in the previous equations gives $\alpha_f = 10.24$, thus $x_1 = 0.5943$, $z_1 = 0.8019$ and $\rho = 0.029$. Then in equation 3.17 the coefficient $0.5 f_c x_1 z_1 = 1.978$ N/mm^2. For a 1:2:4 concrete, CP 2007 gives $f_c = 7$ N/mm^2, $f_s = 85$ N/mm^2 and $\alpha_e = 15$, thus $\alpha_f = 12.14$, $x_1 = 0.5527$, $z_1 = 0.8158$, $\rho = 0.0228$ and $0.5 f_c x_1 z_1 = 1.578$ N/mm^2.

The last paragraph did not make use of equation 3.12. This equation is most useful for obtaining x_1 when the section is fully defined.

Example 3.5. A cantilever wall of a shallow rectangular tank contains a 5 m head of water. Design the cross section at the bottom of the wall in accordance with CP 2007.

Distribution of water pressure on wall is triangular, its maximum being $5 \times 10 = 50$ kN/m^2. For 1 m run of wall, bending moment at base of wall $= (50/2) \times 5 \times (5/3) = 208.3$ kN m/m. Let $h =$ wall thickness.

Using a 1:1.6:3.2 mix the permissible tensile concrete stress for designing against cracking is 1.84 N/mm^2, thus $(1 \times h^2/6) = 208.3/1840 \therefore h = 0.824$ m, say 0.8 m as we have ignored the reinforcement. Using 40 mm cover and 20 mm diameter bars, $d = 800 - 50 = 750$ mm. Designing for strength, assume for speed that the permissible stresses of concrete and steel stated previously apply simultaneously, then using the previous formulae, for concrete $M = 1978 \times 1 \times 0.75^2 = 1113$ kN m, this is more than required, for steel $A_s = 208.3/(85\,000 \times 0.8019 \times 0.75)$ m$^2 = 4075$ mm^2.

From *Table 3.2* use 20 mm diameter bars at 75 mm centres. We need to check that the increased I for an uncracked section, due to the steel, makes h satisfactory. Had we taken $h = 0.824$ or more this last check would be unnecessary.

$$I \simeq 1 \times 0.8^3/12 + 0.004\,19 \times 14 \times (0.75 - 0.4)^2 = 0.049\,85\,\text{m}^4$$

$$\therefore M \simeq 1840 \times 0.049\,85/0.4 = 229.3\,\text{kN m which is} > 208.3$$

A more economical method of obtaining the above steel, because of 1113 being much less than 208.3, is

$$K = 208.3/(1 \times 0.75^2)\text{kN/m}^2 = 0.370\,\text{N/mm}^2$$

From *Table 3.3*, take z_1 as 0.885

$$\therefore A_s = 208.3/(85\,000 \times 0.885 \times 0.75)\,\text{m}^2 = 3692\,\text{mm}^2$$

Table 3.2.

No. of bars	Cross-sectional areas of groups of bars, mm^2							
1	28.3	50.3	78.5	113	201	314	491	804
2	56.6	100.5	157.0	226	402	628	982	1608
3	84.9	151.0	236.0	339	603	942	1473	2412
4	113.0	201.0	314.0	452	804	1256	1964	3217
5	142.0	252.0	393.0	565	1005	1570	2455	4020
6	170.0	302.0	471.0	678	1206	1884	2945	4825
7	198.0	352.0	550.0	792	1407	2198	3436	5629
8	226.0	402.0	628.0	905	1608	2514	3927	6434
9	255.0	452.0	707.0	1020	1810	2830	4420	7240
10	283.0	503.0	785.0	1130	2010	3140	4910	8040
d_b, mm	6	8	10	12	16	20	25	32
50	565.0	1005	1570	2262	4020	6284	9818	16 080
75	377.0	670	1047	1508	2680	4190	6547	10 720
100	283.0	503	785	1131	2010	3140	4910	8 040
125	226.0	402	628	904	1608	2514	3927	6 432
150	188.0	335	523	754	1340	2095	3273	5 360
175	162.0	287	449	646	1149	1794	2805	4 594
200	141.0	251	393	565	1005	1571	2455	4 020
250	113.0	201	314	452	804	1257	1964	3 216
300	94.3	168	262	377	670	1047	1636	2 680

Pitch of bars, mm   Cross-sectional areas of bars per metre, mm^2

From *Table 3.2*, spacing of bars $= (3140/3692) \times 100 = 85$ mm.

Take this as our design and check precisely for *h*. *Table 3.4* is as described in Section 3.2.3, using dm units for convenience. Therefore

$$x = 358.8/85.17 = 4.213 \text{ dm} = 0.4213 \text{ m}$$

$$I = 1998 - 85.17 \times 4.213^2 = 486.3 \text{ dm}^4 = 0.048\,63 \text{ m}^4$$

$$M = 1840 \times 0.048\,63/(0.8 - 0.4213) = 236.3 \text{ kN m}$$

which is > 208.3. Design could be recommenced using a slightly thinner wall, but the reinforcement would be increased slightly so the alteration in cost would be fairly insignificant and may be more or less.

The deflection of the top of a container wall like this can be very important, particularly near corners of rectangular tanks, and the above *I* is suitable for use in such calculations because there are more uncracked than cracked sections. Although the value of *h* was determined so that the wall would not crack, there will be some cracks because of shrinkage, temperature changes and small relative settlements—the design mainly ensures that cracks will be few and small.

F

Table 3.3.

K	0.343	0.459	0.584	0.714	0.849	0.990	1.133	1.278	1.427	1.578	1.731	1.884	1.978
z_1	0.898	0.885	0.873	0.862	0.852	0.844	0.836	0.829	0.822	0.816	0.810	0.805	0.802

$K = M/bd^2 \, \text{N/mm}^2, f_s = 85 \, \text{N/mm}^2, \alpha_e = 15$

Table 3.4.

Portion	A	y	Ay	Ay^2	I_n
Concrete	$10 \times 8 = 80$	4	320	1280	$10 \times 8^3/12 = 427$
Steel	$0.3692 \times 14 = 5.17$	7.5	38.8	291	—
Totals	85.17		358.8	1998	

Example 3.6. A slab with $h = 0.8$ m, $d = 0.75$ m, and 20 mm diameter bars at 85 mm centres as tension reinforcement withstands a bending moment of 208.3 kN m/m. Taking $\alpha_e = 15$, determine the stresses in the steel and the extreme fibre of the concrete.

Consider 1 m width of slab. From *Table 3.2*, $A_s = 3140/0.85 = 3694\,mm^2$, thus $\rho = 0.003\,694/(1 \times 0.75) = 0.004\,925$. From equation 3.12

$$x_1 = -0.073\,88 + \sqrt{(0.073\,88^2 + 2 \times 0.073\,88)} = 0.3176$$

From equation 3.16, $z_1 = 1 - 0.3176/3 = 0.8941$. From equations 3.17 and 3.18

$$f_c = 2 \times 208.3/(0.3176 \times 0.8941 \times 1 \times 0.75^2)\,kN/m^2 = 2.608\,N/mm^2$$

$$f_s = 208.3/(0.003\,694 \times 0.8941 \times 0.75)\,kN\,m^2 = 84.09\,N/mm^2$$

(This demonstrates that the final design of Example 3.5 is in order.)

3.3 Elastic theory for shear stresses

From the elastic theory for bending it is possible to compute the distribution of horizontal shear stresses. From classical elastic theory, the shear force is equal to the rate of change of the bending moment along a beam, and for this to occur the beam has to withstand horizontal shearing stresses. The section of a reinforced concrete beam shown in *Figure 3.4(a)* is symmetrical about a vertical axis. The distributions of bending stresses for two sections distance δx apart are shown in *Figure 3.4(b)*, the bending moments causing the distributions being M and $(M + \delta M)$ respectively. The horizontal shear stress will now be determined at AB. The concrete stress on the small element of area $b\delta y$ is given by

$$f_{c1} = (M/I)\,y \tag{3.19}$$

at one section of *Figure 3.4(b)* and at the other section by

$$f_{c1} + \delta f_{c1} = [(M + \delta M)/I]\,y \tag{3.20}$$

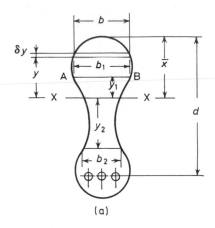

(a)

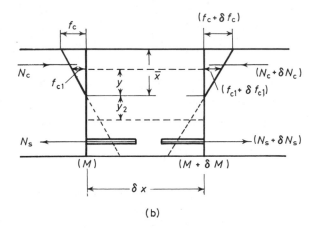

(b)

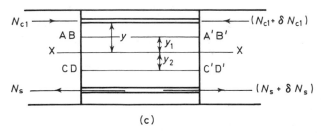

(c)

Fig. 3.4

Subtracting these quantities

$$\delta f_{c1} = (\delta M/I)\, y \tag{3.21}$$

Forces on strip at the two sections are

$$N_{c1} = f_{c1} b \delta y \tag{3.22}$$

$$N_{c1} + \delta N_{c1} = (f_{c1} + \delta f_{c1}) b \delta y \tag{3.23}$$

Subtracting these quantities

$$\delta N_{c1} = \delta f_{c1} b \delta y \tag{3.24}$$

Figure 3.4(c) shows the same two sections as *Figure 3.4(b)*. It can be seen that the plane ABA′B′ has to resist shear stresses due to all such quantities as $(N_{c1} + \delta N_{c1}) - N_{c1} = \delta N_{c1}$. Hence the total shear stress resisted by plane ABA′B′ is given by

$$v = (\Sigma \delta N_{c1})/(b_1 \delta x) \tag{3.25}$$

Substituting from equations 3.24 and 3.21, equation 3.25 becomes

$$v = [\Sigma(\delta M/I)\, yb.\, \delta y]/(b_1 \delta x) \tag{3.26}$$

Now from the well known theory of bending, shear force

$$V = \delta M/\delta x \tag{3.27}$$

Therefore from equations 3.26 and 3.27

$$v = [\Sigma(V\delta x/I)\, yb.\, \delta y]/(b_1 \delta x) = (V/Ib_1) \Sigma by.\, \delta y \tag{3.28}$$

Or more precisely

$$v = (V/Ib_1) \int_{y_1}^{\bar{x}} by.\, dy \tag{3.29}$$

This is the horizontal shearing stress at a point distance y_1 from the neutral axis XX. From classical theory of elasticity it is also the vertical shearing stress at this point. Equation 3.29 has been derived considering the rate of change of compressive stress in the concrete along the beam, and only concerns sections above the neutral axis. Considering a plane CDC′D′ below the neutral axis, the horizontal shear stress resisted by this plane considering forces below it is given by

$$v = [(N_s + \delta N_s) - N_s]/(b_2 \delta x) = (1/b_2)(\delta N_s/\delta x) \tag{3.30}$$

Now from Section 3.2.4, $M = N_s z$ and combining this with equation 3.27

$$V = \delta M/\delta x = z(\delta N_s/\delta x) \tag{3.31}$$

From equations 3.30 and 3.31

$$v = V/zb_2 \tag{3.32}$$

This equation is independent of y_2, hence the shear stress (vertical or horizontal) is constant below the neutral axis.

Equations 3.29 and 3.32 are expressions which apply to any section which is singly reinforced and symmetrical about its vertical axis. Applying these to a rectangular section as shown in *Figure 3.2*, $b_1 = b_2 = b$. Equation 3.29 therefore becomes

$$v = (V/I) \int_{y_1}^{\bar{x}} y \, dy = [V/(2I)] [(\bar{x})^2 - y_1^2] \qquad (3.32a)$$

This gives a parabolic distribution of stress above the neutral axis and the maximum value is at the neutral axis when $y_1 = 0$, thus

$$\text{max. } v = V(\bar{x})^2/(2I) \qquad (3.33)$$

Now from equations of Sections 3.2.3 and 3.2.4,

$$M = N_c z = (f_c/2) \, \bar{x} b z \qquad (3.34)$$

$$M = f_c(I/\bar{x}) \qquad (3.35)$$

Eliminating M between equations 3.34 and 3.35

$$(\bar{x})^2/(2I) = 1/(bz) \qquad (3.36)$$

Substituting this in equation 3.33

$$\text{max. } v = V/zb \qquad (3.37)$$

Below the neutral axis, applying equation 3.32,

$$v = V/(zb) \qquad (3.38)$$

The distribution of shear stress is therefore as shown in *Figure 3.2(e)*. As concrete is much stronger in compression and shear than it is in tension, the principal tensile stresses, often known as the *diagonal tensile stresses*, are the criterion as regards failure due to shearing forces. If the principal tensile stresses due to combining the stresses shown in *Figures 3.2(d)* and *(e)* are computed, below the neutral axis, there are no longitudinal concrete stresses in the diagram. As the horizontal shear stresses by classical theory have equal complementary vertical shear stresses, these combine to give principal diagonal tensile stresses at 45° to the horizontal and equal in magnitude to the horizontal shear stresses. Above the neutral axis the longitudinal compressive stresses reduce the diagonal tensile stresses resulting from combining complementary horizontal and vertical shear stresses. Diagonal tensile stresses help shrinkage stresses in causing cracking. This diagonal cracking is sometimes simultaneous with shear failure for a beam with no web reinforcement.

For T-beams and beams with compression reinforcement, at and below the neutral axis the above applies, that is the maximum diagonal tensile stress is constant and equal to $V/(zb)$.

Example 3.7. From the previous discussion the distribution of horizontal (or vertical) shear stress in the concrete for the section in *Figure 3.3(a)* of Example 3.4 is as shown in *Figure 3.3(b)*, being parabolic for GH, JK and LM. Determine the shear stresses represented by points H, J, K, L, M and N, if the shear force is 60 kN.

For points M and N, i.e. maximum at neutral axis XX, using equation 3.29 (and figures from Example 3.4) it is simpler to consider the section below the neutral axis for the moment of area term.

$$v = [0.06/(0.016\,34 \times 0.16)] \times 0.044\,18 \times (0.847 - 0.3179)\ \text{MN/m}^2$$
$$= 0.5365\ \text{N/mm}^2$$

As explained previously this is equal to $V/(zb)$. Thus alternatively from equation 3.10

$$z = 1.634 \times 10^6/[441.8(84.7 - 31.79)]\ \text{cm} = 699\ \text{mm}$$

then

$$v = 60\,000/(699 \times 160) = 0.5365\ \text{N/mm}^2$$

For approximate preliminary design one would perhaps have guessed that the centre of compression was at about half the depth of the T-flange, giving $z = 847 - 75 = 772$ mm, about 10% error on the dangerous side. The following shear stresses are not needed by the designer but are of academic interest.

For point *H*, using equation 3.29

$$v = \frac{0.06}{0.016\,34 \times 0.45} \times 0.45 \times 0.05\,(0.3179 - 0.025)\ \text{MN/m}^2$$
$$= 0.053\,78\ \text{N/mm}^2$$

For point *J*

$$v = 0.053\,78 + \frac{0.06}{0.016\,34 \times 0.45} \times 0.017\,59 \times (0.3179 - 0.05)$$
$$= 0.053\,78 + 0.038\,45$$
$$= 0.092\,23\ \text{N/mm}^2$$

For point K

$$v = \frac{0.06}{0.016\,34 \times 0.45} \times 0.45 \times 0.15 \times (0.3179 - 0.075) + 0.038\,45$$
$$= 0.1722\ \text{N/mm}^2$$

For point L, $v = 0.1722 \times 0.45/0.16 = 0.4843\ \text{N/mm}^2$

3.4 Shear reinforcement

Generally speaking experimental research[1] shows that if design is based on ultimate strength in shear with suitable load factors, then diagonal crack widths at working loads are acceptable. The ultimate shear forces carried by beams with plain webs have been substituted, by researchers, in equation 3.38 to obtain ultimate values for v. The latter have varied with the many possible variables. Of these variables CP 110 has selected the percentage of longitudinal reinforcement $100A_s/(bd)$ (where d = effective depth) as the most important. CP 110, Table 5, gives ultimate values of $V/(bd)$. It has made the simplification of assuming the lever arm as a constant. As the science is not very accurate this is a not unreasonable assumption. When $V/(bd)$ exceeds these values shear reinforcement must be provided to carry the excess shear force. However, except where $V/(bd)$ is less than half these values, CP 110 requires nominal links to be provided throughout the span[1,2] so that for mild steel links $A_{sv}/s_v = 0.002b_t$, where A_{sv} is the cross-sectional area of the two legs of a link, b_t is the breadth of the beam at the level of the tension reinforcement, and s_v is the spacing of the links $\not> 0.75d$. According to Ref. 4 this should be no different for high yield steel.

For say long continuous beams where temperature stresses assist shrinkage and diagonal tensile stresses, for want of research to the contrary, the writer[2,4] would suggest always using the above nominal links throughout the spans.

No matter how much shear reinforcement is provided, $V/(bd)$ must not exceed the values of Table 6 of CP 110, because steel resists diagonal tension but not the diagonal compression.

Shear reinforcement can be links and/or inclined bars. CP 110 favours a truss analogy method for designing these and adding their ultimate strength to the ultimate shear strength of the concrete from its Table 5. Research shows that beams do not act in this way (e.g. cracks prior to failure are inconsistent with it) but that the ultimate strength design is conservative with this method.

3.4.1 Design of shear reinforcement by CP110 truss analogy

The CP 110 truss analogy method has been judged conservative by research chiefly concerned with vertical stirrups, and stirrups[1] and bars inclined at 45° to the horizontal. Some work with reinforcement at 30° to the horizontal also supports the method. Outside this range one should seek experimental justification. In practice most stirrups are vertical and most bars inclined at 45°.

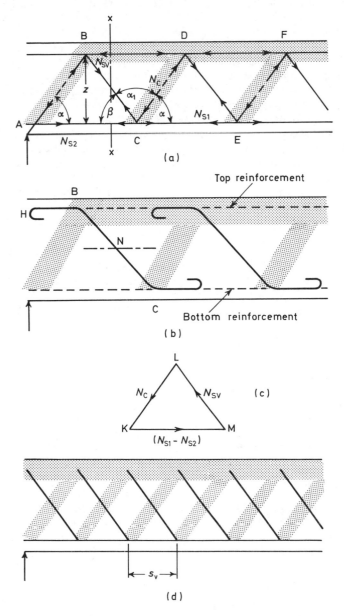

Fig. 3.5

Bars belonging to the main tensile reinforcement are bent up at points such as C and E in *Figure 3.5(a)*. Alternatively independent shear bars (or stirrups) may be used as shown in *Figure 3.5(b)*. A beam is considered to be a statically determinate truss as illustrated in *Figure 3.5(a)*. The longitudinal tension reinforcement is analogous to tension members such as AC and CE in *Figure 3.5(a)*; the concrete resisting longitudinal compression (due to bending) is analogous to compression members such as BD and DF; the bent-up bars are analogous to inclined tension members such as BC and DE, and the inclined compression members such as AB, CD and EF, required to complete the truss analogy, are provided by the concrete of the web. The forces in the analogous truss members AC, BC, DC and EC are as shown, namely N_{s2}, N_{sv}, N_c and N_{s1} respectively. A vector diagram is drawn for these forces in *Figure 3.5(c)*; as the bending moment increases for sections further away from the supports, N_{s1} will be greater than N_{s2} and their difference is represented by the vector KM; forces N_c and N_{sv} are represented by the vectors LK and LM respectively. If the area of tensile reinforcement which is analogous to member CE is A_s, and the area of the bars bent up is ψA_s, and if the bent up bars are required to develop their full stress f_{sv}, then $N_{sv} = \psi A_s f_{sv}$. At the same time, if the stresses in the members CA and CE are not to exceed f_s, they are designed so that $N_{s1} = A_s f_s$ and $N_{s2} = (A_s - \alpha A_s) f_s$. Hence, referring to *Figure 3.5(c)* the vector $LM = \psi A_s f_{sv}$ and the vector $KM = A_s f_s - (A_s - \psi A_s) f_s = \psi A_s f_s$. In the case of mild steel reinforcement, $f_{sv} = f_s$ and therefore $LM = KM$; consequently in the vector diagram LKM,

$$\alpha = \alpha_1 \tag{3.39}$$

For high yield steel[4], using $f_{sv} = 250 \text{ N/mm}^2$ (CP 110 would say 425) and f_s = say 460 N/mm^2; hence $LM = 250 \psi A_s$ and $KM = 460 \psi A_s$, and from the vector diagram

$$\sin \alpha / \sin \alpha_1 = LM/KM = 250/460 \tag{3.40}$$

The inclined compression members are assumed to be sufficiently strong for all requirements. They are safeguarded by compliance with Table 6 of CP 110. By Ritter's Method of Sections, assume the truss to be cut at the section xx shown in *Figure 3.5(a)*. Then resolving vertically for, say, the left-hand side of this section

$$N_{sv} \sin \beta = \text{Shear force at xx} = V \tag{3.41}$$

The principle of the superposition of trusses can be applied. For example, the system shown in *Figure 3.5(d)*, where $s_v = AC/2$, is assumed to be twice as strong as the system of *Figure 3.5(a)*; hence from equation 3.41

$$V = 2N_{sv} \sin \beta \qquad (3.42)$$

The inclined bars shown in *Figures 3.5(a)* and *(d)* are sometimes described as being in *single-shear* and *double-shear* respectively. Extending this principle of superposition for any value of s_v in *Figure 3.5(d)*, equation 3.42 becomes

$$V = (AC/s_v) \, N_{sv} \sin \beta \qquad (3.43)$$

From triangle ABC

$$AC = z \, (\cot \alpha + \cot \beta) \qquad (3.44)$$

This is traditional international truss analogy, but CP110 says $z = d$

$$\therefore AC = d(\cot \alpha + \cot \beta) \qquad (3.44a)$$

Hence equation 3.43 becomes

$$V = (N_{sv}d/s_v) \sin \beta \, (\cot \alpha + \cot \beta) \qquad (3.45)$$

Applying equation 3.45 to mild steel reinforcement and hence using the equation 3.39, also from triangle KLM in *Figure 3.5(c)*,

$$\alpha + \alpha_1 + \beta = 180° \qquad (3.46)$$

Therefore from equation 3.39

$$\alpha = 90° - (\beta/2) \qquad (3.47)$$

Substituting this in equation 3.45

$$V = (N_{sv}d/s_v) \sin \beta \, [\tan(\beta/2) + \cot \beta]$$
$$= (N_{sv}d/s_v) \, [2 \sin^2 (\beta/2) + \cos \beta]$$
$$\therefore V = N_{sv}d/s_v \qquad (3.48)$$

Applying equation 3.45 to high tensile reinforcement, and hence using equation 3.40

$$\sin \alpha = (250/460) \sin \alpha_1$$

Therefore from equation 3.46

$$\sin \alpha = (250/460) \sin (180° - \beta - \alpha)$$
$$\therefore 1.84 \sin \alpha = \sin \beta \cos \alpha + \cos \beta \sin \alpha$$
$$\therefore \cot \alpha = (1.84 - \cos \beta)/\sin \beta \qquad (3.49)$$

Substituting this in equation 3.45

$$V = (N_{sv}d/s_v) \, (1.84 - \cos \beta + \cos \beta) = 1.84 \, (N_{sv}d/s_v) \quad (3.50)$$

Table 3.5.

f_{yv}, N/mm²	d_b, mm	s_v, mm														
		300	275	250	225	200	175	150	125	100	90	80	75	70	60	50
250	6	41.0	44.7	49.2	54.6	61.5	70.3	82.0	98.4	123.0	136.6	153.7	164.0	175.7	205.0	246.0
	8	72.9	79.5	87.5	97.2	109.3	124.9	145.7	174.9	218.6	242.9	273.3	291.5	312.0	364.0	437.0
	10	113.9	124.2	136.7	151.8	170.8	195.2	227.7	273.3	341.6	379.6	427.0	455.5	488.0	569.0	683.0
410	6	67.2	73.3	80.7	89.6	100.8	115.2	134.5	161.3	201.7	224.0	252.0	268.9	288.0	336.0	403.4
	8	119.5	130.4	143.4	159.4	179.3	204.9	239.0	286.9	358.6	398.4	448.0	478.0	512.0	597.0	717.0
	10	186.8	203.7	224.0	249.0	280.0	320.0	373.5	448.0	560.0	622.5	700.0	747.0	800.0	933.8	1120.0

Values of V/d, two-arm stirrups, N/mm, $1/\gamma_m = 0.87$

For inclined bars CP 110 recommends the truss analogy as described, but using $\alpha \not< 45°$. For stirrups CP 110 assumes that $x = d$ and $\alpha = 45°$, so that equation 3.45 becomes

$$V = (N_{sv}d/s_v)(\sin \beta + \cos \beta) \tag{3.51}$$

From equation 3.51 and substituting $N_{sv} = A_{sv}f_{yv}/\gamma_m = A_{sv} 0.87 f_{yv}$

$$A_{sv}/s_v = V/[0.87 f_{yv}(\sin \beta + \cos \beta) d] \tag{3.52}$$

For vertical stirrups $\beta = 90°$, thus

$$A_{sv}/s_v = V/(0.87 f_{yv}d) \tag{3.53}$$

Table 3.5 (upper half) is useful for designers, uses equation 3.53, and refers to mild steel stirrups with $f_{yv} = 250 \text{ N/mm}^2$, from *Table 2.6*. According to Ref. 4 this also applies to all other stirrups. However, for those who wish to use CP 110 for cold deformed hot rolled high yield steel stirrups, $f_{yv} = 410 \text{ N/mm}^2$, the lower half of *Table 3.5* is provided. This would be reasonable also for the use of deformed cold worked high yield steel stirrups, because CP 110 limits f_{yv} to 425 (only 3.7% more than 410).

Table 3.6.

f_{yv}	d_b, mm	10	12	16	20	25	32
		Single bars in single shear at 45°, $1/\gamma_m = 0.87$					
250	V, kN	12.07	17.38	30.91	48.32	75.50	123.7
410	V, kN	19.80	28.51	50.70	79.21	123.8	202.8

For mild steel bars bent up at 45°, from equation 3.39, $\alpha = \alpha_1 = 67.5°$. *Table 3.6* gives shear resistances for single-shear systems for single bars, using equation 3.41 and $1/\gamma_m = 0.87$. Ref. 4 would use $f_{vy} = 250 \text{ N/mm}^2$ for all other bars. CP 110 allows $f_{vy} = 410\text{–}425$ N/mm^2 for deformed high yield steel bars and *Table 3.6* gives shear resistances for $f_{vy} = 410 \text{ N/mm}^2$ which is all right for all deformed high yield bars.

Example 3.8. A beam of T-section has a rib of breadth 250 mm, $d = 600$ mm and $100 A_s/(bd) = 1.2$. Design links to resist an ultimate shear force of 200 kN if the characteristic strength of concrete $= 25 \text{ N/mm}^2$.

$V/(bd) = 0.2/(0.25 \times 0.6) \text{ MN/m}^2 = 1.333 \text{ N/mm}^2$, which is satisfactory from Table 6 of CP 110 (see Section 3.4). From Table 5 of CP 110 (see Section 3.4) shear resistance provided by concrete web alone = [0.65

$+ (1.2 - 1.0) (0.85 - 0.65)] \times 250 \times 600 \text{ N} = 103.5 \text{ kN}$. Hence shear reinforcement is required and it has to resist $200 - 103.5 = 96.5 \text{ kN}$. Using stirrups the V/d required is $96.5/0.6 \text{ kN/m} = 160.8 \text{ N/mm}$. From *Table 3.5* use 6 mm diameter mild steel two-arm stirrups at 75 mm centres ($164 > 160.8$).

Example 3.9. A beam of rectangular cross section has $b = 300 \text{ mm}$, $d = 700 \text{ mm}$, and $100 A_s/(bd) = 1.87$. The ultimate shear force it has to resist is 642 kN. Design a suitable shear reinforcement system. Assume characteristic strength of concrete in compression $= 20 \text{ N/mm}$.

$V/bd = 0.642/(0.3 \times 0.7) \text{ MN/m}^2 = 3.06 \text{ N/mm}^2$, which is satisfactory from Table 6 of CP 110 (see Section 3.4).

From Table 5 of CP 110 (see Section 3.4) shear resistance provided by concrete web alone $= [0.8 + (0.8 - 0.6) \times (1.87 - 1.0)] \times 300 \times 700 \text{ N} = 204.5 \text{ kN}$. Hence shear reinforcement is required and it has to resist $642 - 204.5 = 437.5 \text{ kN}$. According to CP 110 the shear force taken by bent-up bars must not exceed $0.5 \times 437.5 = 218.8 \text{ kN}$. Using pairs of 20 mm diameter bent-up mild steel bars in double shear, from *Table 3.6* this is worth $48.29 \times 4 = 193.2 \text{ kN}$ (< 218.8). Thus the amount to be resisted by stirrups is $437.5 - 193.2 = 244.3 \text{ kN}$, giving $V/d = 244.3/700 \text{ kN/mm} = 349 \text{ N/mm}$. Using mild steel links with two arms, from *Table 3.5*, 10 mm diameter links at 90 mm centres gives $V/d = 379.6 \text{ N/mm}$ (> 349).

3.5 Bond stresses due to shear (or flexural bond)

The theory expounded concerning shear stresses (Section 3.3) assumes perfect adhesion of the concrete to the tensile reinforcement, and therefore involves bond stresses being developed between the steel and the concrete. Referring to *Figure 3.4(b)*, the change of force in the tensile reinforcement between the sections shown is $(N_s + \delta N_s) - N_s = \delta N_s$. This can only be resisted by bond stresses which act on the contact area between the steel and the concrete of $\delta x \Sigma u_s$. Hence the bond stress at this locality is given by

$$f_{bs} = \delta N_s/(\delta x \Sigma u_s) = (1/\Sigma u_s) (dN_s/dx) \qquad (3.54)$$

where $\Sigma u_s = $ sum of the perimeters of bars of tensile steel. Now

$$V = dM/dx = (d/dx) (N_s z) = z(dN_s/dx) \qquad (3.55)$$

Hence from equations 3.54 and 3.55

$$f_{bs} = V/(z \Sigma u_s) = V/(z_1 d\Sigma u_s) \qquad (3.56)$$

These bond stresses are known as *local bond stresses* and ultimate values of $V/(d\Sigma u_s) = z_1 f_{bs}$ are recommended for various types of concrete in Table 21 of CP 110, even though f_{bs} is derived from the elastic theory. Research on ultimate values has been related to $V/(z\Sigma u_s)$, however, and as the results are not very precise it is not unreasonable for CP 110 to take z_1 as constant. Designs need to ensure that ultimate local bond stresses are nowhere exceeded and this is the only requirement in this connection; such bond stresses are local effects and do not for instance require any anchorage.

Example 3.10. The maximum tensile reinforcement in a beam consists of four 25 mm diameter plain bars, and $d = 600$ mm. The maximum ultimate shear force immediately adjacent to a support is 140 kN. If the ultimate local bond stress of Table 21 of CP 110 is 2 N/mm² ($= z_1 f_{bs}$), what is the least number of the reinforcement bars which must continue through to the support? Note that CP 110 calls our $z_1 f_{bs}$ just f_{bs}.

Applying equation 3.56, $\Sigma u_s = 140\,000/(2 \times 600) = 116.7$ mm. The circumference of one 25 mm diameter bar $= \pi \times 25 = 78.5$ mm. Number of bars required to continue through to support $= 116.7/78.5 = 2$, to nearest integer.

3.6 Torsion

Torques are usually calculated assuming a structure to be elastic and uncracked. This is true neither at working nor at ultimate loads, but there is no reliable alternative to this procedure. The monolithic nature of *in situ* construction means that most sections inevitably experience torques, even if only very small, at some time or other. The experience of the designer usually enables him to provide for minor torques when detailing the reinforcement. For example, the external beams to a floor might be given nominal stirruping of say 10 mm diameter at 230 mm centres, as opposed to say 6 mm diameter at 300 mm centres for the internal beams (assuming the possibility of torques on the internal beams is negligible, i.e. a low ratio of live to dead load). This practice is obviously satisfactory in that torsional failures are extremely rare, yet the majority of structures are never overloaded and have been designed to more conservative past codes. CP 110 indicates that where torsional resistance of members can be ignored in analysis of an indeterminate structure, only nominal shear reinforcement (Section 3.4) is required for torsion. If torsional resistance needs assessing, CP 110 requires the torsional rigidity, $G \times C$, of a member to be such that $G = 0.4E_c$ and C, the torsional moment of inertia, equal to half polar second moment of area based on the gross concrete sections. This makes some allowance for the fact that plane cross sections warp under torsion, and the classical theory assumes plane sections remain plane. Torsion failures are very inconsistent and this leads to divergent views upon design by various researchers. In practice, torques often occur simultaneously with shear forces and bending moments, thus complicating the problem still further, especially as the design of members in shear is a difficult problem in itself. In this respect it is good practice to create structural systems so that torsion is always a subsidiary and negligible effect.

Design has been based on the classical work of St. Vernant[5] modified in the light of experimentation. The maximum shear stress due to torsion for a rectangular section is at the middle of the longer sides[5] according to St. Vernant, whereas CP 110 assumes a plastic stress distribution, i.e. a uniform shear stress given by

$$v_t = 6T/[h_{min}^2(3h_{max} - h_{min})] \tag{3.57}$$

where T is the torsional moment due to ultimate loads, h_{min} is the smaller dimension of the section, h_{max} is the larger dimension of the section.

T-, L- or I-sections may be treated by dividing them into their component rectangles, so as to maximise the function $\Sigma(h_{min}^3 h_{max})$ which will generally be achieved if the widest rectangle is made as long as possible. The torsion shear stress carried by each component rectangle can be calculated by treating them as rectangular sections subjected to a torsional moment of

$$T[h_{min}^3 h_{max}/(\Sigma h_{min}^3 h_{max})]$$

Where the torsion shear stress, v_t, exceeds the value $v_{t\,min}$ from Table 7 of CP 110, reinforcement should be provided. In no case should the sum of the shear stresses resulting from shear force and torsion $(v + v_t)$ exceed the value v_{tu} from Table 7 of CP 110 nor, in the case of small sections $(y_1 < 550 \text{ mm})$, should the torsion shear stress, v_t, exceed $v_{tu}y_1/550$, where y_1 is the larger dimension of a link in mm.

Torsion reinforcement should consist of rectangular closed links together with longitudinal reinforcement. CP 110 requires this reinforcement to be additional to any requirements for shear or bending and to be such that:

$$0.87 f_{yv}(A_{sv}/s_v) \geqslant T/(0.8\,x_1y_1) \tag{3.58}$$

$$A_{sl} \geqslant (A_{sv}/s_v)(f_{yv}/f_{yl})(x_1 + y_1) = [T/(0.8\,x_1y_1)][(x_1 + y_1)/0.87f_{y1}] \tag{3.59}$$

where A_{sv} is the area of the legs of closed links at a section, A_{sl} is the area of longitudinal reinforcement, f_{yv} is the characteristic strength of the links, f_{yl} is the characteristic strength of the longitudinal reinforcement, s_v is the spacing of the links, x_1 is the smaller dimension of the links, y_1 is the larger dimension of the links.

In the above formulae f_{yv} and f_{yl} are not to be taken as greater than 425 N/mm². (Ref. 4 would say 250 N/mm².)

Example 3.11. Design links for the section shown in *Figure 3.2*, $h_{max}=488$ mm, to resist an ultimate torsional moment of 3 kN m combined with an ultimate vertical shear force of 60 kN. Concrete is of Grade 25, cover is 25 mm ($x_1 = 100$ mm, and $y_1 = 438$ mm), and $f_{yv} = 250$ N/mm². If $f_{yl} = 425$ N/mm², what extra longitudinal reinforcement is required?

From equation 3.57, $v_t = (6 \times 0.003)/[0.15^2(3 \times 0.488 - 0.15)]$ MN/m² $= 0.6088$ N/mm². From Table 7 of CP 110, this is >0.33 so that torsional reinforcement is required.

$$V/(bd) = 0.06/(0.15 \times 0.45) \text{ MN/m}^2 = 0.8889 \text{ N/mm}^2$$

$$v_t + V/(bd) = 1.498$$

This is in order, as Table 7 of CP 110 limits this to 3.75.

As $y_1 < 550$ mm, v_t must not exceed $3.75 \times 438/550 = 2.99$ N/mm², which is all right as $v_t = 0.6088$ N/mm².

From equation 3.58, $0.87 f_{yv}(A_{sv}/s_v) = 3/(0.8 \times 0.1 \times 0.438)$ kN/m $= 85.62$ N/mm. Using *Table 3.2*, $100A_s/(bd) = (100 \times 982)/(150 \times 450) = 1.455$. From Table 5 of CP 110, $V/(bd) = 0.65 + 0.2 \times 0.455 = 0.741$.

Hence shear reinforcement (two-arm links) is required to resist a value of $V/(bd) = 0.8889 - 0.741 = 0.1479$ N/mm².

Also from Table 6 of CP 110, $0.741 < 3.75$ and is therefore satisfactory.

$$V/d = 0.1479 \times b = 0.1479 \times 150 = 22.19 \text{ N/mm}$$

$$\text{Total } V/d = 85.62 \text{ (see equation 3.53)} + 22.19 = 107.8$$

From *Table 3.5*, use 8 mm diameter two-arm links at 200 mm centres. From equation 3.59, $A_{sl} = 85.62 (100 + 438)/(0.87 \times 425) = 124.6$ mm². Refer to CP 110, as $y_1 > 300$ mm, use two bars in the top corners of the stirrups, two at half depth (of y_1) of stirrup (wired to inside of stirrup) and two in the bottom corners of the stirrups. The latter cannot be catered for by just increasing the size of the tension steel in this case, as the cover would be inadequate. Neither can a bar be placed between these tension bars because of the spacing required between bars (assuming $h_{agg} = 19$ mm). The bottom two bars for torsion will therefore be placed above the tension bars; a clear distance of $19 \times 2/3 = 13$ mm above them. Thus, using *Table 3.2*, six 6 mm diameter bars will be used as the longitudinal torsion bars.

Example 3.12. An ell-shaped beam has: depth and overall breadth of top flange 120 mm and 300 mm respectively, thickness and overall depth of web 100 mm and 600 mm respectively. The ultimate vertical and horizontal shear forces are 20 kN and 10 kN respectively and the ultimate torque is 2 kN m. Determine the reinforcement required for resisting shear and torsion. Concrete is of Grade 30. Cover to longitudinal steel is 20 mm.

Taking the gross web as one rectangle, $\Sigma h_{min}^3 h_{max} = 1^3 \times 6 + 1.2^3 \times (3 - 1) = 9.46$ dm⁴. Taking the gross flange as one rectangle, $\Sigma h_{min}^3 h_{max} = 1.2^3 \times 3 + 1^3 \times (6 - 1.2) = 5.184 + 4.8 = 9.984$ dm⁴. Hence the latter is the way to consider the section as two rectangles.

For the gross flange, torque $= 2 \times 5.184/9.984 = 1.038$ kN m.

G

For the web, torque $= 2 - 1.038 = 0.962$ kN m.
From equation 3.57:

for gross flange $v_t = \dfrac{6 \times 0.001\,038}{0.12^2(3 \times 0.3 - 0.12)}$ MN/m^2 $= 0.5545$ N/mm^2

for web $v_t = \dfrac{6 \times 0.000\,962}{0.1^2(3 \times 0.48 - 0.1)}$ MN/m^2 $= 0.4307$ N/mm^2

From Table 7 of CP 110, these are >0.37 so that torsional reinforcement is required for both gross flange and web.

For gross flange $V/(bd) = 0.01/(0.12 \times 0.27) = 0.3086$ N/mm^2 (assuming $d = 300 - 30 = 270$ mm).

For web $V/(bd) = 0.02/(0.1 \times 0.57) = 0.351$ N/mm^2 (assuming $d = 600 -30 = 570$ mm).

For gross flange $v_t + V/(bd) = 0.8631$.

For web $v_t + V/(bd) = 0.782$

These are in order as Table 7 of CP 110 limits this value to 4.1.

For gross flange $y_1 = 300 - 40 = 260$ mm.

For web $y_1 = 600 - 40 = 560$ mm.

As $y_1 < 550$ for the gross flange, v_t for it must not exceed $4.1 \times 260/550 = 1.938$ N/mm^2, which is all right as $v_t = 0.5545$ N/mm^2.

The design is continued, treating the gross flange and web respectively as in Example 3.11, as though each were independent members.

3.7 Plastic analysis

A material is in a plastic condition when stresses cause permanent deformations, that is when stress is no longer directly proportional to strain (as in Hooke's law). A section of a beam experiences such conditions when realising its ultimate moment of resistance. The plastic method of design predicts the ultimate moment of resistance, and this is required to equal the ultimate bending moment derived from the working loads multiplied by suitable load factors, called the *design loads* by CP 110.

3.7.1 Assumptions of plastic design methods

Plastic design concerns two ideas. Firstly, with regard to the assessment of the bending moments in a redundant frame, plasticity is the ability of highly stressed sections to what might be termed yield, and allow a redistribution[6] of the bending moments towards failure. Secondly, plastic design can be employed in the design of individual sections of structural members. In the latter instance the following assumptions are employed.

It is assumed that plane sections subjected to bending remain plane after bending, which means that the distribution of strain is linear. Some relationship is then assumed between this strain, and stress. This is where the methods differ. Concrete is assumed to have no resistance in flexural tension, perfect bond is assumed between the steel and the concrete, the depth of the steel reinforcement is assumed to be small compared with its effective depth, and normally temperature and shrinkage stresses are ignored in the stress analysis of sections.

3.7.2 Plastic design in bending

The term *balanced design* refers to the situation when the beam is designed to fail simultaneously in flexural compression and tension. *Under-reinforced* sections will fail in flexural tension and *over-reinforced* sections will fail in flexural compression. An under-reinforced section fails owing to yielding (or straining excessively in the case of high yield steel) of the tensile reinforcement; this causes the cracks to open so that the depth of the beam available to resist flexural compression is reduced, and final collapse occurs by the crushing of the compression zone. This is not, however, a *flexural compression failure*, since the failure has actually been precipitated by the inadequacy of the tensile reinforcement and the final failure in apparent *flexural compression* is a secondary effect; it could be described as part of the disintegration of the beam after failure.

Figure 2.6 shows a typical relationship between stress and strain for concrete in compression. As described in Section 2.3.15, this will vary in shape according to the speed of loading, the strength of the concrete, etc. Considerable plasticity is experienced towards failure, i.e. stress is not linearly proportional to strain near failure. It is assumed that the distribution of strain due to bending is linear. The strain is therefore proportional to the distance from the neutral axis. Curves such as those illustrated in *Figure 2.6* can therefore be plotted on the axes *Of* and *Oy* as shown in *Figure 3.6*. For example *Figure 3.6(a)* illustrates the elastic stress distribution at working loads at a section where there is a crack. For higher loads the stress distribution becomes as shown in *Figure 3.6(b)*, and just before failure the stress distribution will be as shown in *Figure 3.6(c)*. The point denoted by g is at the same position on all of *Figures 2.6, 3.6(a), (b)* and *(c)*. Different scales are used for the strains plotted on the axes *Oy*. The diagrams *ehgO* in *Figure 3.6* are termed *stress blocks*.

For estimating the ultimate moments of resistance of beams, the shape of the stress block just before failure must be known. This is assessed empirically, and shapes suggested for the stress block just before failure have included parabolas, cubic parabolas, trapeziums, ellipses, and many unusual shapes; some theories have even assumed that part of the concrete just below the neutral axis resists tensile stresses. This idea is not justified by experiments, because the cracks penetrate too far into the compression zones at the critical sections.

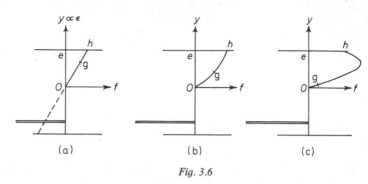

Fig. 3.6

C. S. Whitney, in 1937, suggested considering the stress block as equivalent to a rectangular shape. This leads to a simple theory which has often been found to be more accurate than other methods, e.g. see Ref. 7.

3.7.3 Plastic design of under-reinforced rectangular sections

The distribution of stress at failure is shown in *Figure 3.7*. A general shape is considered for the stress block, the average compressive stress of which is equal to f_{cm}, and the centroid is at a depth of $k_2 x$. Equating longitudinal forces, $N_c = N_s$

$$f_{cm} x b = A_s f_s$$
$$\therefore x = A_s f_s / (f_{cm} b) \tag{3.60}$$

Taking moments about the line of action of N_c the ultimate resistance moment

$$M_u = N_s z = N_s (d - k_2 x) \tag{3.61}$$

Substituting for x from equation 3.60 this becomes

$$M_u = N_s [d - k_2 A_s f_s / (f_{cm} b)] \tag{3.62}$$
$$\therefore M_u = A_s f_s d [1 - (k_2 \rho f_s / f_{cm})]$$
$$= f_s \rho b d^2 [1 - (k_2 \rho f_s / f_{cm})] \tag{3.63}$$

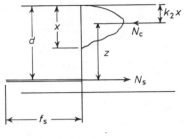

Fig. 3.7

where $\rho = A_s/(bd)$. Whitney and the simplified method of CP110 use a rectangular stress block such that $f_{cm} = 0.85\ f_c'$ (where $f_c' = $ U.S.A. cylinder strength $\simeq 0.84 f_{cu}$) and $0.4 f_{cu}$ respectively, $k_2 = 0.5$ for both, and f_s is f_y for Whitney and f_y/γ_m for CP110 where $\gamma_m = 1.15$. With the equivalent (unlike actual) stress block of Whitney the depth of the stress block x_1 is less than the depth of the neutral axis x. The above equations would use x_1 instead of x in this instance. Whitney gives a good prediction of how a beam will actually fail[7]. The coefficients quoted for Whitney's theory in this chapter assume $f_{cu} \leqslant$

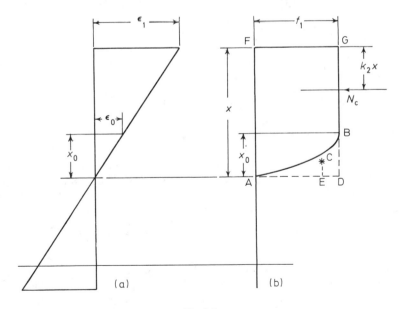

Fig. 3.8

33.33 N/mm². For higher values of f_{cu} refer to Section 8.4.5. The simplified rectangular stress block of CP 110 is chosen to have $x_1 = x$. CP 110 gives a reliably conservative prediction of failure, distorted to ensure that flexural tension rather than compression failures will occur. The former failure gives plenty of warning—large deflections and cracks before failure—whereas the latter failure is very sudden.

The method claimed by CP 110 to be more precise than its simplified method uses a stress block as shown in *Figure 3.8(b)* and the distribution of strain shown in *Figure 3.8(a)*, where $\varepsilon_1 = 0.0035$. Tests over many years show that the maximum extreme fibre compressive strain realised before failure in flexure is about this figure. CP 110 specifies $\varepsilon_0 = \{\sqrt{f_{cu}}\}/5000, f_1 = 0.45 f_{cu}$ and curve AB as a parabola. Thus, considering the shape ABD, its area is AD × BD/3, C is its centroid and CE = BD/4. The compression force N_c is

$$f_{cm}xb = (\text{area ABGF})b$$

$$\therefore f_{cm}x = \text{area ADGF} - \text{area ADB} = f_1 x - f_1 x_0/3$$

$$\therefore f_{cm} = f_1[1 - x_0/(3x)] = f_1[1 - \varepsilon_0/(3\varepsilon_1)]$$

$$\therefore f_{cm} = 0.45 f_{cu}[1 - \{\sqrt{(f_{cu})}\}/52.5] \tag{3.64}$$

Taking moments for compression force about F

$$N_c k_2 x = b[(\text{area ADGF})\,0.5x - (\text{area ADB})(x - \text{CE})]$$

$$f_{cm}xk_2x = 0.5f_1 x^2 - (f_1 x_0/3)(x - x_0/4)$$

$$\therefore k_2 = (f_1/f_{cm})[0.5 - [x_0/(3x)]\{1 - x_0/(4x)\}]$$

$$= \{0.45f_{cu}/(2f_{cm})\}[1 - \{2\varepsilon_0/(3\varepsilon_1)\}\{1 - \varepsilon_0/(4\varepsilon_1)\}]$$

$$\therefore k_2 = (0.225f_{cu}/f_{cm})[1 - \{(\sqrt{f_{cu}})/26.25\}[1 - (\sqrt{f_{cu}})/70]] \tag{3.65}$$

Equations 3.64 and 3.65 are the same as given on page v, Appendix A of Part 2 of CP 110, and are the basis of the design charts.

3.7.4 Balanced plastic design of rectangular sections

The equations of Section 3.7.3 apply. With these equations, as A_s increases x increases and M_u increases, but experimentally we find that x cannot increase beyond a certain amount and increasing the reinforcement further gives no increase in M_u, the section being known as *over-reinforced*. When x has its maximum value, and A_s corresponds to this, the section is in its balanced design condition, the maximum flexural compression being balanced by the minimum A_s to give a maximum M_u for the section.

For balanced design Whitney gives $x_1 = 0.537d$, and CP 110, for design purposes, gives $x = 0.5d$. Using the simplified CP 110 method from Section 3.7.3, equation 3.60 becomes

$$0.5d = A_s f_s/(0.4f_{cu}b), \therefore \rho = 0.2(f_{cu}/f_s) \tag{3.66}$$

Equation 3.61 becomes

$$M_u = A_s f_s(d - 0.5 \times 0.5d) = 0.75 A_s f_s d \tag{3.67}$$

and substituting for ρf_s from equation 3.66

$$M_u = 0.75 \, \rho f_s bd^2 = 0.15 f_{cu} bd^2 = K_1 bd^2 \tag{3.68}$$

Table 3.7.

f_y,	f_s,	f_{cu}, N/mm²					
N/mm²	N/mm²	20	25	30	40	50	
250	217	1.843	2.304	2.765	3.687	4.608	
410	357	1.120	1.401	1.681	2.241	2.801	
460	400	1.000	1.250	1.500	2.000	2.500	ρ,
425	370	1.081	1.351	1.622	2.162	2.703	%
485	422	0.948	1.185	1.422	1.896	2.370	
K_1,	N/mm²	3.0	3.75	4.5	6.0	7.5	

Equations 3.66 and 3.68 are used for design *Table 3.7*. Without tables, equation 3.68 is usually used to decide the size of the member as limited by the strength of the concrete. Then A_s is often obtained from equation 3.67 thus:

$$A_s = \frac{M_u}{0.75df_s} = \frac{M_u \gamma_m}{0.75df_y} = \frac{1.15M_u}{0.75df_y} = \frac{1.533M_u}{df_y} \tag{3.69}$$

Example 3.13. A slab 160 mm thick is reinforced in tension with 16 mm diameter bars having 30 mm cover. Determine the spacing of the reinforcement if the slab is designed in accordance with CP 110 for an ultimate resistance moment of 27.6 kN m, and if $f_{cu} = 25$ N/mm² $= 25000$ kN/m², $f_y = 250$ N/mm² $= 250000$ kN/m² and γ_m for the steel $= 1.15$.

Using simplified CP 110 method, from equation 3.68 considering 1 m width of slab, for balanced design

$$M_u = 0.15 \times 25000 \times 1 \times (0.160 - 0.038)^2 = 0.15 \times 25000 \times 0.122^2$$
$$= 55.82 \text{ kN m}$$

This is greater than 27.6 kN m, hence section is under-reinforced. From equation 3.62 (or 3.63), using $f_s = 250000/1.15 = 217400\,\text{kN/m}^2$

$$27.6 = 217400\,A_s\,[0.122 - 0.5 \times 217400 A_s/(0.4 \times 25000 \times 1)]$$

$$\therefore A_s = 0.001\,161\,\text{m}^2 = 1161\,\text{mm}^2$$

From *Table 3.2*, use 16 mm diameter bars at 150 mm centres.

Example 3.14. Repeat Example 3.13 using the method preferred by CP 110. Equations 3.64 and 3.65 give

$$f_{cm} = 0.45 \times 25\,[1 - \{(\sqrt{25})/52.5\}] = 10.18\,\text{N/mm}^2$$

$$k_2 = (0.225 \times 25/10.18)\,[\![1 - [\{(\sqrt{25})/26.25\}\{1 - (\sqrt{25})/70\}]]\!] = 0.4548$$

From equations 3.60 and 3.62 for 1 m width of slab for balanced design

$$0.5 \times 122 = A_s 217.4/(10.18 \times 1000)$$

$$M_u = A_s \times 217.4\,[122 - 0.4548 A_s 217.4/(10.18 \times 1000)]$$

$$\therefore M_u = 61 \times 10180\,(122 - 0.4548 \times 61)\,\text{N mm} = 58.53\,\text{kN m}$$

This is >27.6, hence section is under-reinforced. Hence from equation 3.62

$$27.6 = 217400 A_s\,[0.122 - 0.4548 \times 217400 A_s/(10180 \times 1)]$$

$$\therefore A_s = 0.001\,145\,\text{m}^2 = 1145\,\text{mm}^2$$

From *Table 3.2*, use 16 mm diameter bars at 175 mm centres. (Page ix of CP 110, Part 2, obtains the same answer by using the design charts.)

Example 3.15. The slab of Example 3.13 is reinforced in flexural tension with 16 mm diameter bars at 175 mm centres, i.e. $A_s = 1149\,\text{mm}^2$ per metre and is to be tested to destruction. Predict its ultimate resistance moment using Whitney's theory[7].

To determine whether it is under- or over-reinforced, apply equation 3.60 (referring also to Sections 3.7.3 and 3.7.4).

$$x_1 = [1149 \times 250/\{0.85(0.84 \times 25) \times 1000\}] = 16.09\,\text{mm}$$

For balanced design $x_1 = 0.537 \times 122 = 65.5\,\text{mm}$, hence section is under-reinforced. Hence applying equation 3.62

$$M_u = 1149 \times 250\,[122 - 0.5 \times 1149 \times 250/\{0.85(0.84 \times 25) \times 1000\}]$$

$$\text{N mm} = 32.73\,\text{kN m}$$

This is considerably greater than the 27.6 kN m used in Example 3.14, indicating the conservativeness built into the CP 110 design method.

Example 3.16. Design a section of a beam, using the simplified CP 110 method, to have an ultimate resistance moment of 200 kN m, using $f_{cu} = 20\,\text{N/mm}^2$, $f_y = 250\,\text{N/mm}^2$ and γ_m for steel = 1.15.

From equation 3.68, $200 \times 10^6 = 0.15 \times 20\,bd^2$, $\therefore bd^2 = 66.67 \times 10^6$.

If $b \simeq 0.5d$ say, then $d^3 = 133.3 \times 10^6$ and $d = 511$. So $b = 255$, say, choose $b = 250$ mm. Then $d = \sqrt{(66.67 \times 10^6/250)} = 516$ mm. From equation 3.69 (or 3.66 or 3.67)

$$A_s = 1.533 \times 200 \times 10^6/(516 \times 250) = 2377 \text{ mm}^2$$

From *Table 3.2* use three 32 mm diameter bars.

Example 3.17. Repeat Example 3.16 using *Table 3.7*.
From *Table 3.7*, $K_1 = 3 \text{ N/mm}^2$ and $\rho = 1.843\%$. Using equation 3.68, $bd^2 = 200 \times 10^6/3 = 66.67 \times 10^6$. As in Example 3.15, choose $b = 250$ mm, then $d = 516$ mm. Then $A_s = 0.018\,43 \times 250 \times 516 = 2377 \text{ mm}^2$. From *Table 3.2* use three 32 mm diameter bars.

3.7.5 Plastic design of any shape of under-reinforced section

For the section of *Figure 3.4(a)*, using a rectangular concrete stress block of average stress f_{cm} (see Section 3.7.3), equating longitudinal forces

$$f_{cm}A_c = A_s f_s \tag{3.70}$$

where A_c = area concrete in compression. Taking moments about the line of action of N_c

$$M_u = A_s f_s z \tag{3.71}$$

where z = lever arm = distance between lines of action of N_c and N_s. N_c acts at centroid of A_c.
Whitney specifies $f_{cm} = 0.85 f_c' \approx 0.85 \times 0.84 f_{cu} = 0.714 f_{cu}$ as before. CP 110 specifies $f_{cm} = 0.4 f_{cu}$ for simplified design method.

3.7.6 Balanced plastic design of any shape of section

For balanced design (see Section 3.7.4) the depth of the stress block x_1 obtained from equation 3.70 is $0.537d$ for Whitney's theory and $0.5d$ for CP 110.

Example 3.18. A T-beam has a flange of breadth 750 mm and depth 130 mm. The width of its rib or web is 300 mm and the tensile reinforcement comprises one layer of five 25 mm diameter bars having an effective depth of 500 mm. Determine its ultimate resistance moment from the simplified design method of CP 110, assuming $f_{cu} = 20 \text{ N/mm}^2$, $f_y = 425 \text{ N/mm}^2$ and $\gamma_m = 1.15$ for the reinforcement.
From equation 3.70 and *Table 3.2*, $0.4 \times 20[300x + (750 - 300) \times 130] = 2455 \times (425/1.15)$, $\therefore x = 183$ mm.
As $0.5 \times 500 = 250$ the section is under-reinforced. Also $x >$ depth of flange, hence beam is designed as a T-beam and not a rectangular beam.

If depth of centroid of A_c is $k_2 x$ then taking area moments about the top of the beam for A_c

$$A_c k_2 x = (300x^2/2) + (750 - 300) \times (130^2/2)$$

$$\therefore k_2 x = [8\,826\,000/\{300x + (750 - 300) \times 130\}] = 77.8 \text{ mm}$$

From equation 3.71

$$M_u = 2455 \times (425/1.15) \times (500 - 77.8) \text{ N mm} = 383.1 \text{ kN m}$$

3.7.7 Plastic design of any shape of under-reinforced section containing compression steel

It might be said that compression reinforcement is only required in a beam when the balanced design condition applies. Whilst this is often true, there are cases where compression steel is available even though not required to assist flexural compression, e.g. sometimes at the supports of continuous beams. In such cases the compression steel can increase the ultimate bending moment of the section and sometimes economises in tensile steel.

For a section like *Figure 3.4(a)* but including compression steel in the top, using a rectangular concrete stress block (see Section 3.7.3):

Compression force for gross area of concrete in compression = $A_c f_{cm}$

Compression force for compression steel over and above that included at this position above = $A'_s(f_{sc} - f_{cm})$

Therefore equating longitudinal forces

$$A_c f_{cm} + A'_s(f_{sc} - f_{cm}) = A_s f_s \tag{3.72}$$

where A_s = gross area concrete in compression, A'_s = area of compression steel and f_{sc} = stress in compression steel (usually characteristic strength because the strain is high in the concrete and thus the steel as flexural concrete failure occurs). Taking moments about the line of action of N_c

$$M_u = A_s f_s z_1 = A_s f_s(d - k_2 x_1) \tag{3.73}$$

Whitney specifies $f_{cm} \simeq 0.714 f_{cu}$ as before. CP 110 specifies $f_{cm} = 0.4 f_{cu}$ for simplified design method. Whitney gives f_{sc} as yield stress of compression steel and CP 110 gives f_{sc} as $2000 f_y/(2000\gamma_m + f_y)$, where $\gamma_m = 1.15$, which it suggests can be simplified to $0.72 f_y$ for ease of calculation. There is no need to make this simplification if use is made of *Table 3.8*. These comments on f_{sc} depend upon the strain in the compression steel being at least that corresponding to its

Table 3.8

f_y, N/mm²	250	410	460	425	485
$2000 f_y/(2300 + f_y)$, N/mm²	196.1	302.6	333.3	311.9	348.3
$0.72 f_y$, N/mm²	180.0	295.2	331.2	306.0	349.2

yield stress. The strain at the level of the compression steel needs to be assessed and related to the stress–strain relationship for the steel (e.g. CP 110, Fig. 2)—see Section 3.7.10.

3.7.8 Balanced plastic design for any shape of section containing compression steel

For balanced design (see Sections 3.7.4 and 3.7.6) the depth of the stress block x_1 is $0.537d$ for Whitney's theory and $0.5d$ for CP 110.

Example 3.19. Determine the ultimate resistance moment from the simplified design method of CP 110 of the beam section shown in *Figure 3.3*, where the reinforcement bars are 10 mm diameter in compression and 32 mm diameter in tension and have 40 mm cover of concrete. Assume $f_s = 250/1.15 = 217.4$ N/mm², $f_{sc} = 196.1$ N/mm², and $f_{cu} = 0.4 \times 25 = 10$ N/mm².

From equation 3.72 and *Table 3.2* $[160x + (450 - 160) \times 150] \times 10 + 314(196.1 - 10) = 4825 \times 217.4$, $\therefore x = 347.2$.

This is > 150, hence beam is designed as a T- and not a rectangular beam. For balanced design, whether T- or rectangular section, $x = 0.5 \times (900 - 56) = 422$ mm. Hence section is under-reinforced. Taking moments about top of beam for compression forces

$$k_2 x[\{160 \times 347.2 + (450 - 160) \times 150\} \times 10 + 314(196.1 - 10)]$$

$$= [160 \times (347.2^2/2) + (450 - 160) \times (150^2/2)] \times 10 + 314 \times (196.1 - 10)$$

$$\times 45$$

$$\therefore k_2 x = 125.6 \text{ mm}$$

From equation 3.73, $M_u = 4825 \times 217.4 \times [(900 - 56) - 125.6]$ N mm $= 753.6$ kN m. This assumes that the compression steel is not near the bottom of the stress block. Effective depth of compression steel $= d' = 45$ mm, whereas $x = 347.2$. From CP 110 (see Section 3.7.10) this matters when $d' > 0.2143d$. In this example d' is much less than $0.2d$. If the compression steel is near the neutral axis (rather an unusual case) refer to Section 3.7.10.

3.7.9 Design of compression steel for rectangular section

In practice the commonest place where compression steel is required is at the supports of continuous *in situ* T-beams. The bending moments at mid span and supports are of similar magnitude; the T-section at mid span enables the rib (or stem) there to be small compared to the size of a rectangular beam; then at the support the bending moment is reversed and the beam is designed as a rectangular beam, with the small rib as its compression zone. Under these circumstances the section here may require compression steel. Thus a rectangular section has to be designed to take a bending moment in excess of its balanced design bending moment by the addition of compression steel.

Example 3.20. Design a rectangular section 300 mm wide by 600 mm deep to have an ultimate resistance moment of 300 kN m in accordance with CP 110. Assume $f_{cu} = 20$ N/mm^2, $f_y = 250$ N/mm^2 and γ_m for steel = 1.15.

For balanced design (with no compression steel) see Section 3.7.4, and applying equation 3.68, estimating $d = 560$ mm, $M_u = 3 \times 300 \times 560^2$ N mm = 282.2 kN m.

Hence section needs compression steel. An estimate of $d' = 35$ mm. Then z for compression steel = $d - d' = 525$ mm and z for concrete in compression = $0.75 \times 560 = 420$ mm, because depth of stress block is 0.5×560. Thus, using *Table 3.8*

$$A'_s = [(300 - 282.2) \times 10^6/(196.1 \times 525)] = 172.9 \text{ mm}^2$$

From *Table 3.2* use say two 12 mm diameter bars. Resolving forces longitudinally (i.e. using equation 3.72), $(250/1.15) \times A_s = 300 \times 0.5 \times 560 \times 0.4 \times 20 + 172.9 \times (196.1 - 0.4 \times 20)$. $A_s = 3241$ mm^2.

From *Table 3.2* use say seven 25 mm diameter bars. These will need to be in two layers, say five in the bottom and two in the layer above. Using 19 mm down coarse aggregate the vertical distance between the layers of bars = 13 mm, say 15 mm. This will mean that, using 25 mm cover to the tension steel, an estimate of $d \approx 550$ mm (the accurate value necessitates the calculation of the position of the centroid of these bars). Using 25 mm cover for the compression reinforcement, $d' = 31$ mm. This design can be repeated with these more accurate values of d and d', but it should not alter the results as the reinforcement is on the generous side because of the limitation of bar sizes, and the initial estimates of d and d' were not too inaccurate. $d' \not> 0.2143d$, hence (see Section 3.7.10) the stress we have taken in the compression steel does not need reducing.

3.7.10 Compression steel near to neutral axis

In practice this can hardly ever arise, as when compression steel is required it is placed as far from the neutral axis as possible for economic reasons. At failure in flexure the maximum strain in the concrete is about 0.0035 and the distribution of strain is approximately linear.

Hence the strain at the level of the compression steel is 0.0035 $(x - d')/x$. According to Fig. 2 of CP 110, if this strain is less than 0.002 then the stress–strain curve of Fig. 2 should be used to determine the design stress in the compression reinforcement. Hence for CP 110 we do not have to concern ourselves in Sections 3.7.7, 3.7.8 and 3.7.9 with reducing the stress in the compression steel if $(x - d')/x \not< 20/35$, i.e. $x \not< 2.333d'$. In the case of balanced design $x = 0.5d$, this becomes $0.5d \not< 2.333d'$, i.e. $d' \not> 0.2143d$ (CP 110 calls this $0.2d$).

3.7.11 Further points about compression steel

Compression steel, even if available in a section, should not be relied upon in design if not prevented by adequate anchoring from buckling out of the faces of the member; each bar should be anchored at right-angles to the outer surface of the concrete according to CP 114, but CP 110 has reduced this requirement in its Clause 3.11.4.3. Both codes specify diameter and spacing of suitable stirrups. For example, framing bars in a beam are not always suitably anchored for compression steel when evaluating ultimate resistance moment.

Compression steel, even if available in a section, should not be relied upon in design without adequate compression laps. For example, steel in the bottom of a continuous T-beam over a support with nominal lapping can only be used to the strength of the lapping in compression.

3.8 Limit state of deflection

Deflections can be calculated as in Example 3.2. This assumes the gross concrete section to be homogeneous and the deflection is obtained with elastic theory. The value assumed for E_c or α_e (as $E_c = E_s/\alpha_e$) can vary considerably (see Section 2.3.15). For accurate work it is best to obtain E_c from laboratory tests on specimens of the concrete. In loading tests on *in situ* buildings with say Grade 20 (CP 110) concrete perhaps about 2–3 months old, the writer has experienced α_e of about 10, i.e. due to the live load applied. In design it is useful to divorce the live and dead loadings and take $\alpha_e = 8$ for strong concretes to 10 for weak concretes for calculating deflections due to live loads (i.e. of short duration; not developing much creep), and take $\alpha_e = 15$ for deflections due to dead loading (this will be realised over several years of creep).

Ignoring the reinforcement and including concrete in tension, which at the positions of cracks will not exist, is usual practice. In the writer's experience troubles with deflection arising from design are usually due to no calculations of deflections, on at least these lines, being made. In the laboratory, obtaining E_c and E_s from tests of specimens of the concrete and steel respectively, and allowing concrete to take tensile stresses and allowing for reinforcement to obtain I, deflections of beams can be predicted very accurately[1] before cracks about 0.01 mm wide occur. For greater loads the deflection often approaches the deflection calculated in the same way but excluding concrete in tension. Just before failure it often becomes greater than this calculated amount.

For a beam (span l) carrying uniformly distributed loading q and if the breadth is a constant proportion of its depth and if E_c is constant, then maximum deflection $\sim ql^4/bd^3 \sim ql^4/d^4$. Thus the l/d ratio can be a guide to deflection, but only in conjunction with q. The tables restricting l/d in CP 114 for beams and slabs were inadequate in that q was ignored. Tables 8 and 9 of CP 110 are similar but require modification by factors given in Tables 10 and 11, but whose derivation and justification are not given. For example, for a constant l/d the greater q the greater the deflection (even though the reinforcement will be increased slightly). Now from Table 10 the greater the reinforcement the less the factor, which reduces the allowable l/d ratio, so indirectly some allowance is made for q. Table 11 has similar logic but also allows for the fact that when compression steel is present it restrains the tendency of the shrinkage in this location to increase deflections.

Deflections must be limited so as not to cause trouble to internal partitions and finishes. Beams obviously sagging are aesthetically undesirable—the deflection can be calculated and the beam given an upward camber of at least this amount. Slightly hogging beams are aesthetically acceptable. Consideration should be given to each particular case, and CP 110 gives general guidance on limitation of deflection.

3.9 Limit state of cracking

Research has indicated that water cannot penetrate to the reinforcement to cause corrosion if cracks are not greater than 0.25 mm wide. This figure can vary with the concrete grade, cover, etc., and CP 110 uses a figure of 0.3 mm, specifying other figures for various exposures. CP 110 considers that its reinforcement detailing recommendations take care of undesirable cracking. For example, smaller diameter bars at closer centres resist cracks much better than the converse. When this problem is of particular importance because, say, of severe exposure, or where groups of bars are used, an empirical formula is given in Appendix A of CP 110 for assessing crack widths.

REFERENCES

1. Wilby, C. B., *Strength of Reinforced Concrete Beams in Shear*, PhD Thesis, University of Leeds (1949)
2. Wilby, C. B., 'Permissible Shear Stresses of the 1957 British Code of Practice', *Journ. Amer. Conc. Inst.*, June (1958)
3. 'Lessons from Failures of Concrete Structures', *A.C.I. Monograph No. 1* (1965)
4. Wilby, C. B., see Ref. 9 of Chapter 2
5. Evans, R. H., and Wilby, C. B., *Concrete—Plain, Reinforced, Prestressed and Shell*, Ed. Arnold (1963)
6. Wilby, C. B., and Pandit, C. P., 'Inelastic Behaviour of Reinforced Concrete Single Bay Portal Frames', *Civil Eng. and Pub. Wks. Rev.*, Mar. (1967)
7. Evans, R. H., 'The Plastic Theories for the Ultimate Strength of Reinforced Concrete Beams', *Journ. Inst. Civil Engrs.*, Dec. (1943)

Chapter 4

Reinforced concrete slabs

4.1 Slabs spanning one-way

These are designed per unit width as rectangular beams. See examples in Chapter 3.

4.2 Slabs spanning two-ways

These are, for example, *in situ* rectangular slabs supported on four, three or two adjacent sides. Originally they were designed by ascertaining bending moments and shear forces by elastic theory and then designing sections for these by elastic theory (Sections 3.2.4 and 3.3). Subsequently it was possible (CP 114) alternatively to design the resistance to bending moments by plastic theory (Section 3.7.2). This seems to have been satisfactory, but, of course, is very illogical, as towards failure the distribution of the bending moments will be different to that given by the elastic theory.

Bending moments from elastic theory can be calculated from simple formulae in CP 110 for rectangular slabs carrying uniformly distributed loads. These and formulae for slabs with triangularly distributed loadings (for walls of tanks) and with concentrated loads are given in *Reynolds' Handbook*.

The latest step has been to design slabs by assessing the bending moments at collapse by Johansen's yield line[1] or Hillerborg's strip method[2].

Generally shear stresses are low and usually found to be satisfactory when checked. Slab thicknesses are often dictated by deflection considerations and sometimes slabs have to have a minimum practical thickness of preferably 125 mm. Deflections may be calculated from elastic theory but are more simply dealt with using Tables 8, 9 and 10 of CP 110. Cracks can be controlled at working loads by attention to detailing (see CP 110).

4.3 Flat slabs

These are slabs without beams supported only by columns. Flared column heads usefully reduce the high shear stresses in the slabs around the column heads. Flat slabs generally give a heavier construction than beam and slab systems; they require more concrete and steel but the shuttering is much less expensive. For longer spans of flat slabs *dropped panels* are sometimes used to make the construction lighter in weight. This usually means dropping the soffits of rectangular portions of slab around the column heads.

Design has been based on empirical formulae which are limited to systems with rectangular panels, length-to-width not exceeding 4/3, with at least three continuous spans in both directions. Such formulae are given in CP 110 and are simple to use. The alternative method allowed by CP 110 is more arduous but is useful when the empirical formulae do not apply. It consists of dividing the structure longitudinally and transversely into frames consisting of columns and the connecting strips of the slabs, and then elastically analysing these frames for bending moments and shear forces. This is well enunciated in CP 110. More recently they might be designed using Johansen's yield line[1] or Hillerborg's strip method[2].

4.4 Yield line theory of slab analysis

There are two different methods of analysis, perhaps most simply explained by the following two examples.

Example 4.1. A square *isotropically* reinforced slab (this means the slab is reinforced identically in orthogonal directions, which means that its ultimate resisting moment is the same in these two directions and along any line in any other direction—see proof in Section 4.4.1) is simply supported along all of its sides. Determine by the *equilibrium method of analysis* the ultimate

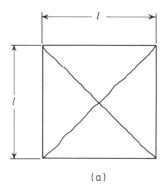

(a)

(b)

Fig. 4.1

H

resisting moment m_u per unit length of yield line balancing an ultimate uniformly distributed load q kN/m^2 (this includes the self-weight of the slab).

It is easy to imagine that the slab will essentially fail by the diagonals of *Figure 4.1(a)* becoming yield lines. That is, cracks will occur along these lines in the soffit of the slab and they will open as the tensile steel yields. Steel can maintain its yield stress as the steel yields, so the section rotates for no increase in moment, but eventually the rotation becomes excessive (extreme fibre strain $\simeq 0.0035$) and the concrete compression zone disintegrates. As the rotation at the centre of each yield line becomes considerable, but not excessive, due to yielding of the steel there, the rotation near the corners of each yield line eventually becomes sufficient for the steel to have also yielded there. Failure is precipitated therefore when each unit of length of each yield line has reached its ultimate bending strength. Generally the rotation at the centre of a yield line will not have been sufficient to cause failure there before the ultimate bending moments near the ends of the yield line have been realised. Thus m_u is constant along each yield line and is the same for each because of symmetry. Considering the equilibrium of the moments of any one of the identical slab segments about the support (see *Figure 4.1(b)*) the total bending moment along AC is $m_u l/\sqrt{2}$, and is shown by a vector such that the moment acts in a clockwise direction when viewed along the vector arrow. This vector has a component parallel to AB of $(1/\sqrt{2})(m_u l/\sqrt{2})$. Similarly for CB. Hence taking moments about AB

$$(ql^2/4)(l/6) = 2(1/\sqrt{2})(m_u l/\sqrt{2})$$

$$\therefore m_u = ql^2/24 \text{ kN m/m}$$

Example 4.2. The rectangular isotropically reinforced slab shown in *Figure 4.2* is simply supported along all of its sides. Determine by *virtual work analysis* the ultimate resisting moment m_u per unit length of yield line balancing a total ultimate uniformly distributed load of q kN/m^2.

It is easy to imagine that the slab will essentially fail by the yield lines shown in *Figure 4.2(a)*. The distance l_1 is unknown but will be such as to maximise the ultimate resistance moment required to balance the ultimate loading. A simple procedure, which lends itself to solution by computer, is by trial and error. In *Figure 4.2(a)* angle ACD is 90°, AC $= \sqrt{(1.5^2 + l_1^2)}$. Triangles CBA and DBC are similar, thus CD:CA = CB:AB and triangles ACE and ACB are similar, thus EC:AB = AC:CB. That is

$$l_2 = (1.5/l_1)\sqrt{(2.25 + l_1^2)} \text{ and } l_3 = (l_1/1.5)\sqrt{(2.25 + l_1^2)}$$

Considering AC, from *Figure 4.2(b)*, the total angle of rotation at this yield line for a small unit increase of deflection at C in radians is

$$\frac{1}{l_3} + \frac{1}{l_2} = \frac{1.5}{l_1\sqrt{(2.25 + l_1^2)}} + \frac{l_1}{1.5\sqrt{(2.25 + l_1^2)}} = \frac{1}{\sqrt{(2.25 + l_1^2)}}\left(\frac{1.5}{l_1} + \frac{l_1}{1.5}\right)$$

Similarly for this unit deflection at C the total rotation of the yield line CF is $1/1.5 + 1/1.5 = 1.333$. For our first trial let $l_1 = 2.1$ m. Then

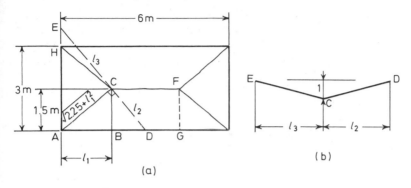

Fig. 4.2

$AC = \sqrt{(2.25 + 4.41)} = 2.58$ m. The rotation at $AC = (1/2.58)(1.5/2.1 + 2.1/1.5) = 0.8195$.

The internal work done (bending moment × angular rotation) as the unit incremental deflection occurs at yield is

$$m_u \times 2.58 \times 0.8195 \times 4 + m_u(6 - 2 \times 2.1) \times 1.333 = 10.86m_u$$

The external work done (making use of symmetry) whilst this incremental deflection occurs is

$$2\,(\text{Load on AHC}) \times \tfrac{1}{3} + 2\,(\text{Load on CFGB}) \times \tfrac{1}{2} + 4\,(\text{Load on ABC}) \times \tfrac{1}{3}$$

$$= 0.667 \times (1.5 \times 2.1q) + (6 - 2 \times 2.1) \times 1.5q + 1.333\,(0.75 \times 2.1q)$$

$$= 6.9q$$

Equating internal and external works done $m_u = (6.9/10.86)q = 0.6354q$.

Trying other values for l_1 and summarising for values of l_1 of 1.8, 1.95, 2.1 and 2.25 the corresponding values of m_u/q are 0.635, 0.637, 0.635 and 0.632 respectively. For a given q the maximum $m_u = 0.637q$ kN m/m corresponding to the yield pattern when $l_1 = 1.95$ m.

4.4.1 Reinforcement

If a slab is not isotropically reinforced (see Example 4.1), its ultimate strengths are different in two perpendicular directions and it is *orthogonally anisotropically* or simply *orthotropically* reinforced. When isotropically reinforced (see Example 4.1), its ultimate resistance moment is the same in any direction. This will now be proved. As the lever arm is assumed constant, for the bending moment per unit length to be constant in any direction it is only necessary to prove that the force provided by the tensile reinforcement per unit length is constant in any direction. Referring to *Figure 4.3* the reinforcement has the same spacing s in each rectilinear direction and the force in each bar is N_s. Considering the line CD the component of N_s at A perpendicular to this line is $N_s \cos \alpha$. Also AB = $s/\cos \alpha$. Thus the force per unit length of CD and perpendicular to CD due to the bars in the direction AE is $(N_s \cos^2 \alpha)/s$. The component of N_s at D perpendicular to CD is $N_s \sin \alpha$. Also CD = $s/\sin \alpha$. Thus the force per unit length of CD and perpendicular to CD due to the bars in the direction DF is $(N_s \sin^2 \alpha)/s$. Thus the total force per unit length of CD and perpendicular to CD is $(\cos^2 \alpha + \sin^2 \alpha) N_s/s = N_s/s$ which is the same as the force per unit length in either of the two rectilinear directions of the reinforcement, Q.E.D.

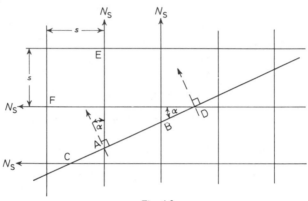

Fig. 4.3

Slabs that are orthotropically reinforced can be dealt with by altering the dimensions for design purposes[3].

In the above examples the sections are assumed to be under-reinforced—this is normally the case for slabs, because of deflection and minimum thickness requirements. The analyses are dependent upon all yield lines being able to develop fully before say the initial portion loses its moment-carrying capacity due to excessive rotation—the extreme fibre strain reaching about 0.0035. The reinforcement per unit length can be different in each of two rectilinear directions, but must be constant along any line, otherwise in the above examples m_u would not be constant along each line. The analysis is most convenient for slabs of difficult shapes and slabs with holes or openings, where an elastic analysis is difficult. R. H. Wood[4] says that a slab designed elastically, stopping off all bars whenever one could, would generally be more economic than if it were designed by yield line (or strip method), and was to be preferred as a design. The reinforcement would not be so simple a system. The yield line analysis offers no information on the best distribution of the steel, but can be used to analyse a slab where the steel has been distributed according to some other method (e.g. Hillerborg's). Curtailing bars means that yield lines have to be considered at sections where the bars are discontinued.

If cracks need to open considerably the bars across the crack tend to kink to endeavour to be at right-angles to the cracks; this gives a slightly stronger resistance moment (up to about 14%). Hence the designer can ignore kinking and his design will be slightly safer.

Membrane action can help the strength of a slab when the deflections are large towards failure. It is reasonable for the designer to ignore it and have slightly extra safety.

4.4.2 Further points on yield line analyses

Both of the previous methods of analysis give what is termed *upper bound* solutions, in that it might always be possible to think of some other pattern of yield lines which might require a greater m_u to balance a given loading. For example, the tank wall shown in

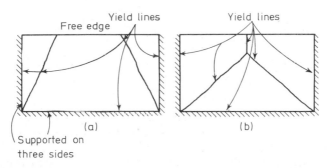

Fig. 4.4

Figure 4.4 may fail by either of the yield line mechanisms shown, and to design the wall both have to be investigated. Essentially there is not much difficulty in choosing various possible sensible yield line patterns and thus in practice there is no great need to worry about this upper bound problem. One chooses from experimental experience of failures or from imagining how failures might occur.

Yield lines are generally straight, lie along lines of encastré supports, pass over columns, and pass through the intersection of rotating adjacent slab elements. Strictly speaking when a yield line meets an unsupported edge it must do so perpendicularly, as yield line moments are maximum moments. However, if the yield line away from this edge is skew and straight, it is usually continued in a straight line to the edge[3]. This makes negligible error in the calculations.

In the previous examples an alternative possibility is for the yield line pattern to be as shown in *Figure 4.5*. If the corners are not held down each corner element such as ABCD will rotate about AC lifting B from the support. If the corners are held down, at each corner, lines such as AC will become yield lines. In this case reinforcement is required perpendicularly to lines such as AC in the top of such a slab. The slab spans between AD and CD, and AB and BC, and sometimes supplementary reinforcement is desired to take care

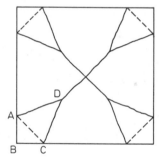

Fig. 4.5

of this, the bars being parallel to the direction AC and in the bottom of the slab. The yield lines at corners are called *corner levers*. Although their effect is adverse they are often neglected in yield line analyses for simplicity. For right-angled corners this causes an error of about 9 % with bottom steel only and much less when there is top as well as bottom steel. The error is particularly high[5] for acute angles with free edges, about 26 % with bottom steel only and 14 % when there is top and bottom steel.

For non-rectangular shapes Ref. 6 is useful.

4.5 Hillerborg's strip method of slab design

This is perhaps most simply explained by the following example.

Example 4.3. Design the slab shown in *Figure 4.6* which has edges restrained against rotation and has to carry an ultimate uniformly distributed total load of 15 kN/m^2.

It seems sensible (based on our experience of elastic theory, or tests) to reduce the reinforcement parallel to supports towards supports from mid span. For simplicity in practice we shall design for six bands of reinforcement in each direction, so we have to divide the loading to give constant loading for the width of each band. In the x-direction, from symmetry we need to design only three bands, and the typical strips to be designed, as representative of each band, are LM, NP, and QR. The only load these

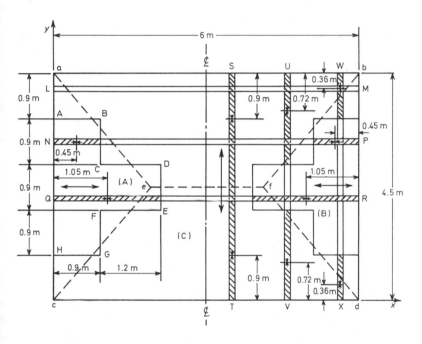

Fig. 4.6

strips are designed to carry is that on their shaded portions, there being none on LM in this instance. Similarly in the y-direction the strips ST, UV and WX are designed to carry the load on their shaded portions. Thus we have chosen that the two zones such as ABCDEFGHA, now called (A) and (B), are to be carried by strips in the x-direction and the remainder of the slab, zone (C), is to be carried by strips in the y-direction. This means that we have chosen that the load on the two zones (A) and (B) is carried by strips to the edges ac and bd, and that the load on the zone (C) is carried by strips to the edges ab and dc. This kind of loading on the edges is in line with much past practice for deciding loads on supporting peripheral beams, and is more recognisable if the internal *discontinuity lines*, such as ABCDEFGH, were ae, ec, bf and fd. The stepped discontinuity lines were chosen to approximate to ae, ec, bf and fd, because of the desire to have the reinforcement in bands. The discontinuity lines are chosen to be a sensible (with regard to one's experience of elastic theory, yield line, and/or experimentation) division of the areas of the slab likely to be carried by each support.

The distribution of bending moments and shear forces in each strip can be determined by either elastic or plastic theory. For example, for strip ST, taking a nominal breadth of 1 m, the total load along ST is $15 \times 4.5 = 67.5$ kN. By elastic theory, it is a fixed beam, so each support moment is $67.5 \times 4.5/12 = 25.31$ kN m and the mid-span moment is $25.31/2 = 12.66$ kN m. By plastic theory, suppose we choose to keep the maximum bending moment to a minimum, i.e. make the support and mid-span moments equal, then either of these $= 67.5 \times 4.5/16 = 18.98$ kN m.

4.5.1 Further points on Hillerborg's strip method

This method involves a tremendous amount of plastic action. If one's experience of elastic analysis, yield line or tests is severely gone against in deciding discontinuity lines and points of contraflexure of strips, then a very undesirable slab can be obtained, with regard to cracking. It is also possible that such a slab might not pass the British Standard loading test because of lack of recovery of deflection due to high plasticity, even though the ultimate strength might be satisfactory.

If the strips are designed elastically then the method is very illogical, in that deflections of strips are not matched up, as was done years ago by Grashof-Rankine.

In the last decade at least, there has been a tendency to design structures more accurately using computers, so the strip method seems a retrograde step in one sense. It is the kind of method one has not been proud to use in design offices when precise elastic analyses have been too difficult to attempt in the time available. But there are more computer packages available today.

The method is illogical, in that load on a portion cannot just be carried in one direction and not other directions, and a prime advantage of plate action is thus lost.

The great advantage of the method is that it is easy to use and apply to any shape. It is probably best when there is not an elastic analysis available.

For skew slabs the strips are taken as beams cranked in plan and the geometry involves different portions of a strip having different widths[4].

It is an easy method to apply to slabs containing holes[4].

4.6 CP 110 and yield line and strip methods

CP 110 recommends that yield line and strip methods can be used provided the ratio between support and span moments does not deviate from elastic theory by more than 50%. This helps to safeguard against a slab which may crack badly at working loads. But it restricts the great advantage of these methods, which is to design slabs which cannot be designed in the time available by elastic methods.

An experienced designer can usually design using yield line and strip methods, knowing that he is not grossly offending serviceability without having also to design by elastic theory.

Certain plastic methods used[2, 4] choose the positions of the points of inflection or contraflexure at 0.2 of the span from each support

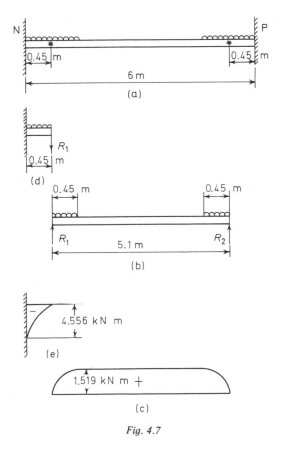

Fig. 4.7

for strips such as ST, 0.4 of the length of the loaded area from each support for strips such as UV and WX, 0.5 of the length of the loaded area from each support for strips such as NP and QR (strip LM having no loading). These points are marked with an asterisk in *Figure 4.6*. On this basis *Figure 4.7(a)* shows the loading on strip NP and its points of contraflexure. *Figure 4.7(b)* shows how the bending moments are to be calculated for the portion of NP between the points of contraflexure. Reaction $R_1 = R_2 = 0.45 \times 15 = 6.75$ kN. The bending moment diagram is shown in *Figure 4.7(c)* and its maximum bending moment is $R_1 \times 0.45 - 0.45 \times 15 \times 0.225 = 1.519$ kN m. *Figure 4.7(d)* shows how the bending moments are to be calculated for the portion of NP between the points of contraflexure and the supports. The bending moment for the portion shown is shown in *Figure 4.7(e)* and its maximum value is $R_1 \times 0.45 + 15 \times 0.45 \times 0.225 = 4.556$ kN m. Shear forces can be calculated correspondingly. Other strips can be treated similarly.

REFERENCES

1. Jones, L. L., and Wood, R. H., *Yield Line Analysis of Slabs*, Thames and Hudson and Chatto and Windus, London (1967)
2. Woods, R. H., and Armer, G. S. T., 'The Theory of the Strip Method for Design of Slabs', *Proc. I.C.E.*, Oct. (1968)
3. Dunham, C. W., *Advanced Reinforced Concrete*, McGraw-Hill, U.S.A. (1964)
4. Ferguson, P. M., *Reinforced Concrete Fundamentals*, John Wiley, U.S.A. (1973)
5. Hughes, B. P., *Limit State Theory for Reinforced Concrete*, Pitman, London (1971)
6. European Committee for Concrete, *Information Bulletin No. 35*, English Translation issued by C. and C.A., London (1962)

Chapter 5

Columns and walls

5.1 General

CP 110 recommends ultimate load design using plastic theories and not elastic theory as was allowed by CP 114.

5.2 Slender columns

Slender and *short* columns are ones affected and not affected by buckling respectively. CP 110 defines a column as short when its 'effective length (height') is less than 12 times its least lateral dimension. A slender column is designed as a short column but requiring it to withstand an additional bending moment given by CP 110 equations 33–38 inclusive. These are empirical formulae based on classical buckling theory.

5.3 Axially loaded short columns

The assumptions of this analysis are given in Section 3.7.1. *Figure 5.1(a)* shows the cross section of a column and *Figure 5.1(b)* the distribution of stress across the cross section. Basically the concrete strains in a plastic fashion until the reinforcement yields (or, for high yield steel, its strain is so great as) to realise the maximum strain which can be tolerated by the concrete. The latter occurs when the stress in the concrete is about $0.67 f_{cu}$. Hence resolving vertically

$$\text{Ultimate axial load} = 0.67 f_{cu} A_c + f'_y A_{sc} \qquad (5.1)$$

where A_c = area of concrete, A_{sc} = area of compression steel, f_{cu} = characteristic strength of concrete and f'_y = characteristic strength of steel in compression. Refer to Section 3.7.7, which explains that for design, CP 110 approximates f'_y/γ_m to $0.72 f_y$. f_{cu} is divided by a γ_m of 1.5 so the ultimate axial load for CP 110 design purposes $= (0.67/1.5) f_{cu} A_c + 0.72 f_y A_{sc} = 0.45 f_{cu} A_c + 0.72 f_y A_{sc}$, where f_y is the characteristic tensile strength of the steel. As loads in practice are rarely axial, to allow for an eccentricity up to $0.05 \times$ least lateral dimension CP 110 recommends for design an ultimate axial load

$$= 0.4 f_{cu} A_c + 0.67 A_{sc} f_y \qquad (5.2)$$

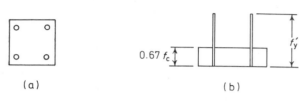

(a)   (b)

Fig. 5.1

Example 5.1. Design a short reinforced concrete column for an ultimate axial load of 2900 kN.

Suppose $f_{cu} = 25$ N/mm² and $f_y = 250$ N/mm² and assume say 2% of reinforcement, then $A_{sc} = 0.02(A_c + A_{sc})$, $\therefore A_{sc} = 0.02041 A_c$. From equation 5.2, $2900 = 0.4 \times 25000 \times A_c + 0.67 \times 250000 \times 0.02041 A_c$, $\therefore A_c = 0.2161$ m². Therefore $A_c = 0.2161$ m² and $A_{sc} = 0.02041 \times 0.2161 = 0.004411$ m². Gross sectional area of column $= 0.2161 + 0.0044 = 0.2205$ m². Use a column 470mm square with (see *Table 3.2*) four 32 mm diameter and four 20 mm diameter bars.

5.4 Plastic analysis for eccentrically loaded short columns

This is a column required to be designed for an ultimate axial load N combined with an ultimate bending moment M where $M = Ne$. *Figure 5.2(a)* shows the cross section of a column of any shape, *Figure 5.2(b)* the distribution of stress assumed by CP 110 for design purposes, and *Figure 5.2(c)* the distribution of strain. The $0.4 f_{cu}$ should really be $(0.67/1.5) f_{cu} = 0.45 f_{cu}$ but this is reduced to $0.4 f_{cu}$ to give slightly less chance of failure as a concrete compression failure is sudden and thus undesirable.

Resolving vertical forces $N = N_c + N_{sc} - N_s$ where N_c is the force in the concrete over the gross area in compression, N_{sc} and N_s are forces in the steel in compression and tension respectively. CP 110 ignores the fact that concrete does not exist over the cross-sectional areas of steel—generally a useful and satisfactory assump-

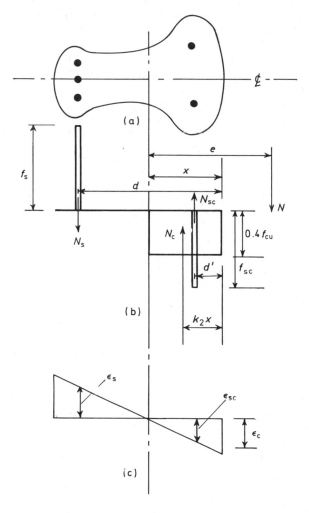

Fig. 5.2

tion. Thus

$$N = 0.4 f_{cu} A_c + A_{sc} f_{sc} - A_s f_s \qquad (5.3)$$

where A_c = area of concrete in compression, A_{sc} and A_s = areas of steel in compression and tension respectively, $k_2 x$ is the distance to the line of action of N_c (i.e. to the centroid of A_c), and f_{sc} and f_c = design strengths (stresses) of compression and tension steels respectively.

Taking moments for convenience about the line of action of N_s

$$N(e + d - x) = N_c(d - k_2x) + N_{sc}(d - d') \qquad (5.4)$$

For large eccentricities, failure is initiated by the tension steel yielding or straining excessively (for high yield steel) causing the value of x to be reduced until eventually the concrete crushes. For small eccentricities the concrete may crush to cause failure when the tension steel is in compression or only modestly strained in tension. Between these two types of failure we have what is called a *balanced condition* where the failure is caused by simultaneous crushing of the concrete and yielding or excessive straining of the tension steel. For this condition let $N = N_b$ and $e = e_b$. Then from *Figure 5.2(c)*, taking $\varepsilon_c = 0.0035$ because tests show that this is approximately the maximum strain which is experienced at crushing of the concrete, and taking $\varepsilon_s = 0.002 + f_s/200$ from CP 110, Fig. 2,

$$x/d = \varepsilon_c/(\varepsilon_c + \varepsilon_s) = 0.0035/(0.0055 + f_s/E_s) \qquad (5.5)$$

where f_s is $f_y/\gamma_m = f_y/1.15$.

For this condition, for a defined section, x is given by equation 5.5, then A_c calculated and then equation 5.3 gives N_b if we know f_{sc}. From *Figure 5.2(c)* ε_{sc} can be found for this balanced condition and, from the stress–strain curve for this steel, f_{sc} can be determined. Normally ε_{sc} is large enough to develop the maximum stress in the steel (it might not do so if this steel were unusually near to the neutral axis). Then e_b can be determined from equation 5.4.

Loads with eccentricities less than e_b cause primary compression failures at ultimate loads greater than N_b, whereas loads with eccentricities greater than e_b cause primary tension failures at loads smaller than N_b.

Thus if $e > e_b$, $f_s = f_y/1.15$ and $\varepsilon_c = 0.0035$. Assume $f_{sc} = 0.72f_y$ or take a more accurate value from *Table 3.8*. Then for a defined section and a known e, equations 5.3 and 5.4 can be solved for the two unknowns N and x.

But if $e < e_b$, $\varepsilon_c = 0.0035$, so equation 5.5 has two unknowns x and f_s. If this f_s is substituted in equations 5.3 and 5.4 (and f_{sc} obtained from *Table 3.8*) and then N eliminated between these two equations, a cubic equation for x results. It may be solved by trial and error (a computer can help), estimating sensible values of x. When x is obtained N can then be obtained from either of the equations from which it was eliminated.

In the first case ε_{sc} and in the second case ε_s and ε_{sc} can be determined finally from *Figure 5.2(c)* to see if they are great enough to correspond to the values of f_{sc} and f_s assumed, using the stress–strain curve of CP 110, Fig. 2. If not, then values of f_{sc} and f_s are estimated and the above calculations repeated until the values assumed for f_{sc} and f_s have values ε_{sc} and ε_s which agree with their values on the stress–strain curves.

In the following examples, eccentricity is specified from the centre line of a column, as this is a more practical case for the reasons given in Section 5.6.

Example 5.2. The cross section of a column is rectangular of width 250 mm ($= b$) by depth 450 mm ($= h$), and $A_s = A_{sc} = 1473\,mm^2$ (three 25 mm diameter bars—see *Table 3.2*), $d' = 50\,mm$ and $d = 450 - 50 = 400\,mm$. If $f_{cu} = 25\,N/mm^2$, f_y for A_s and A_{sc} is 250 N/mm², $E_s = 200\,kN/mm^2$ and the eccentricity of the line of action of the load from the centre line of the column $= e_1 = 420\,mm$, determine the CP 110 ultimate load for the column.

For balanced design condition $f_s = 250/1.15 = 217.4\,N/mm^2$. From equation 5.5

$$x/0.4 = 0.0035/(0.0055 + 0.2174/200) \qquad \therefore x = 0.2125\,m$$

From *Figure 5.2(c)*, $\varepsilon_{sc} = 0.0035 \times (212.5 - 50)/212.5 = 0.002676$. This is > 0.002, so from CP 110, Fig. 2, and *Table 3.8* $f_{sc} = 196.1\,N/mm^2$.

From equation 5.3, $N_b = 0.4 \times 25\,000 \times 0.25 \times 0.2125 + 0.001\,473 \times 196\,100 - 0.001\,473 \times 217\,400 = 531.3 + 288.9 - 320.2 = 500.0\,kN$.

From equation 5.4, $N_b(e_b + 0.4 - 0.2125) = 531.3 \times (0.4 - 0.2125/2) + 288.9 \times 0.35 = 257.2 \therefore e_b = 0.3269\,m$. Therefore value of e_1 for balanced design $= e_b - x + h/2 = 0.3269 - 0.2125 + 0.225 = 0.3394\,m$.

This is less than 420 mm, hence failure is by yielding of tension steel. Equation 5.3 gives

$$N = 0.4 \times 25\,000 \times 0.25x + 288.9 - 320.2 = 2500x - 31.3$$

$$e = e_1 - h/2 + x$$

$$\therefore e + d - x = e_1 - h/2 + d = 0.42 - 0.225 + 0.4 = 0.595\,m.$$

From equation 5.4

$$0.595N = 2500x\,(0.4 - x/2) + 288.9 \times 0.35$$

From the above two equations in N and x, $x = 0.1708$ and $N = 395.7\,kN$.

(It is interesting to note that in *Table 3.8* there is a greater percentage difference in the values of f_{sc} for the lower concrete strengths. In this example if $f_{sc} = 180.0$ is used instead of 196.1 then $x = 0.177\,m$ and $N = 387.5\,kN$.)

Now as $\varepsilon_c = 0.0035$, from *Figure 5.2(c)*

$$\varepsilon_{sc} = 0.0035 \times (170.8 - 50)/170.8 = 0.00248$$

This is greater than 0.002 (see CP 110, Fig. 2), so the value assumed for f_{sc} is correct. (Had this not been so, it would be necessary to obtain f_{sc} from the strain ε_{sc} on CP 110, Fig. 2. Then repeat the above calculations. Then the ε_{sc} calculated would correspond to a slightly different value of f_{sc} to the one taken. The whole process is repeated as many times as necessary to obtain the required accuracy of N).

Example 5.3. The cross section of a column is rectangular and $b = 250$ mm, $h = 450$ mm, $d' = 50$ mm and $d = 450 - 50 = 400$ mm. If $f_{cu} = 25$ N/mm^2, f_y for A_s and A_{sc} is 250 N/mm^2, $E_s = 200$ kN/mm^2, $e_1 = 420$ mm, $N = 395.7$ kN and $A_s = A_{sc}$ determine $(A_s + A_{sc})$ using the CP 110, Part 2, Design Charts.

$d/h = 400/450 = 0.8889$, use Chart 44.

$M/(bh^2) = 0.3957 \times 0.42/(0.25 \times 0.45^2)$ MN/m$^2 = 3.28$ N/mm^2.

$N/(bh) = 0.3957/(0.25 \times 0.45)$ MN/m$^2 = 3.52$ N/mm^2.

$\therefore$ "A_{sc}" $= 2.55 \times 250 \times 450/100 = 2869$ mm^2 ("A_{sc}" in CP 110 $= A_s + A_{sc}$).

For design, Chart 45 rather than 44 would be used to be on the safe side. Chart 44 is used as it is nearer to the truth and it was desired to compare this result with Example 5.2. The steel area of 2869 mm^2 compares with $2 \times 1473 = 2946$ mm^2. The small difference between the Charts and Example 5.2 is due to the fact that (a) the stress blocks are slightly different, (b) the charts do not use the correct value of d/h.

It is interesting to note that linearly interpolating between Charts 44 and 45 gives 2890 mm^2, which is within 2 % of 2946 mm^2.

Example 5.4. Repeat Example 5.2, only using $e_1 = 180$ mm.

As before, the value of e_1 for the balanced design condition $= 0.3394$ m. This is greater than 0.18 m, hence failure is by compression of concrete. Equations 5.3., 5.4 and 5.5 become

$$N = 0.4 \times 25\,000 \times 0.25x + 288.9 - 0.001\,473 f_s \tag{5.6}$$

$$e + d - x = e_1 - h/2 + d = 0.18 - 0.225 + 0.4 = 0.355 \text{ m}$$

$$0.355N = 2500x\,(0.4 - x/2) + 288.9 \times 0.35 \tag{5.7}$$

Assuming f_s is on the first portion of the stress–strain curve of CP 110, Fig. 2, then from *Figure 5.2(c)*

$$x/0.4 = 0.0035/(0.0035 + f_s/200\,000\,000) \tag{5.8}$$

One way of solving these equations is to assume that f_s is say 217 400 kN/m^2 (i.e. 250/1.15 N/mm^2), then calculate x from the last equation. With these values calculate values of N from the previous two equations. These will normally differ. Adjust the value of f_s and start again. Repeat until the values of N from the two equations are sufficiently in agreement. Alternatively the equations can be algebraically reduced to $1250x^3 - 112.5x^2 + 367.4x - 146.4 = 0$. It is then very easy and rapid to solve this with an electronic hand programmable calculator and guessing values of x, or with a computer library program. The former method gave $x = 0.3191m$, $f_s = 177\,500$ kN/m^2, and $N = 825.2$ kN. From *Figure 5.2(c)*

$$\varepsilon_{sc} = 0.0035 \times (319.1 - 50)/319.1 = 0.002\,952$$

This is > 0.002 (see CP 110, Fig. 2), so our assumption for f_{sc} is correct.

Also from *Figure 5.2(c)*

$$\varepsilon_s = 0.0035 \times (400 - 319.1)/319.1 = 0.000\,887\,3$$

Referring to CP 110, Fig. 2, $0.8f_y/\gamma_m = 0.8 \times 217.4 = 173.9$ N/mm^2, and the corresponding strain is $0.1739/200 = 0.000\,869\,5$. Thus the strain in this steel appears to be within the second linear portion of the stress–strain curve. The co-ordinates of two points connected by this line are (0.000 869 5, 173 900) and (0.002, 196 100). Thus for any point on this line

$$(\varepsilon_s - 0.000\,869\,5)/(f_s - 173\,900) = (0.002 - 0.000\,869\,5)/(196\,100 - 173\,900)$$

$$\therefore f_s = 19\,640\,000\varepsilon_s + 156\,800 \qquad (5.8a)$$

Thus the previous assumption that f_s was on the first portion of the stress–strain curve is incorrect and equation 5.8 becomes

$$x/0.4 = 0.0035/[0.0035 + (f_s - 156\,800)/19\,640\,000] \qquad (5.9)$$

From equation 5.8a the value of f_s corresponding to $\varepsilon_s = 0.000\,8873$ can be obtained. The above calculations for N, x, ε_{sc} and ε_s are then repeated. The whole process can be repeated until the value of N has sufficient accuracy. It converges rapidly. To reduce the arithmetic one might like to plot the above mentioned line of the stress–strain curve so that the values of f_s for various values of ε_s can be read graphically. Alternatively, by direct calculation the above equations can be algebraically reduced to $1250x^3 -112.5x^2 - 44.6x - 14.38 = 0$.

It is very easy and rapid to program this on an electronic hand programmable calculator and solve by trial and error as x is known to be near to 0.3191. This gave $x = 0.3170$, $f_s = 174\,800$ kN/m^2, and $N = 823.9$ kN. Had we guessed initially that the second portion of the stress–strain curve was relevant, not the first portion as in equation 5.8, then we would have used equation 5.9 instead of equation 5.8 and saved considerable time and effort. As x is now different to when ε_{sc} was previously checked, ε_{sc} will be rechecked. From *Figure 5.2(c)*

$$\varepsilon_{sc} = 0.0035 \times (317 - 50)/317 = 0.002\,948$$

This is > 0.002 (see CP 110, Fig. 2), so our assumption for f_{sc} is still correct.

Example 5.5. Repeat Example 5.3, only using $e_1 = 180$ mm, and $N = 823.9$ kN. $d/h = 400/450 = 0.889$ (use Chart 44).
$M/(bh^2) = 0.8239 \times 0.18/(0.25 \times 0.45^2)$ MN/m$^2 = 2.929$ N/mm^2.
$N/(bh) = 0.8239/(0.25 \times 0.45)$ MN/m$^2 = 7.324$ N/mm^2.
"A_{sc}"$= 2.2 \times 250 \times 450/100 = 2475$ mm^2.
Comparing this example with Example 5.4, the steel area of 2475 mm^2 compares with 2946 mm^2. The difference is due to the fact that (a) the stress blocks are slightly different, (b) the charts do not use the correct value of d/h.

It is interesting to note that if the preceding analysis is repeated using the more complicated stress block of Fig. 3, CP 110, Part 2, and if linear interpolation is used between the Charts 44 and 45 of CP 110, then the answer given by the calculation is within 1 % of the answer given by the Charts.

I

5.4.1 Design of eccentrically loaded columns

To be in accordance with CP 110 it is probably best to choose columns from the Charts of Parts 2 and 3 of CP 110. If a column section cannot be obtained in these Charts then the Charts can give guidance in estimating approximately the dimensions of, and steel in, the column. Then it has to be checked as in Section 5.4.

5.5 Reinforced concrete walls

Load bearing reinforced concrete walls are designed as columns, but if any structural reliance is made on the reinforcement, such reinforcement needs to have ties across the wall to prevent the bars buckling outwards. Such ties are highly undesirable in practice, causing much trouble to both the steelfixer and concretor. It is therefore usually more economical to design the wall as though it contained no reinforcement. It would not, however, be built without any reinforcement because differential settlement, shrinkage and temperature expansion or contraction could all cause cracking, which would be most noticeable on a concrete surface. Such small movements also cause hair cracks between the bricks of brickwork walls, but even if occasional bricks are cracked the cracks blend with the pattern of the wall and are not noticeable to the layman. Cracks in concrete surfaces tend to concentrate into a few of large size, rather than many of a small size, and ramble in various directions in an unsightly way. Consequently horizontal and vertical reinforcement is placed in both faces of a reinforced concrete wall, whether the wall is load bearing or not, the horizontal reinforcement usually being nearer the surface than the vertical reinforcement. In practice, the vertical bars are usually made of at least 12 mm diameter, except in the case of very thin walls, as these have to support the horizontal reinforcement. The construction of walls may be very difficult if light reinforcement fabrics are used.

5.6 Design of columns to frameworks

In accordance with CP 110, frameworks are analysed using elastic theory for forces and bending moments, assuming the members to be concentrated at their centre lines. The designer may then choose to redistribute these bending moments as described in Chapter 6. Each column section then needs to be designed for a bending moment about its centre line and an axial force whose line of action is through this centre line. It will be appreciated from Section 5.4 and its examples that a direct design calculation is difficult because of decisions as to whether primary compression or tension failures or balanced design conditions are relevant. The designer will often desire the column to be as large as possible to aid detailing of column and interconnecting beam reinforcement, to avoid long column instability, and for economy as the concrete is a more economic material than steel with regard to the carrying of compression forces. However, the larger the column the more it restricts circulation space in the building, and for this reason and aesthetic considerations the architect will often want columns to be as slender as possible. The designer often chooses the size of the column on this basis, and an assessment of strength. To assess the size of the column and its reinforcement to carry the load required a very approximate design is usually made. This can then be checked more accurately by using the design charts of CP 110, or analytically, similar to the method of Section 5.4 if the section is not included in the design charts. If the approximate design is inadequate or uneconomic then this design is altered accordingly and the above procedure repeated until the designer is satisfied. This is a long process if charts, or a computer program, are not used. For the initial approximate design the gross cross-sectional area can be obtained by dividing the ultimate axial load by $0.42 f_{cu}$ if the line of action of the eccentric load is outside and $0.45 f_{cu}$ within the section. These figures are for rectangular or square cross sections. For circular cross sections the figures would be $0.39 f_{cu}$ and $0.42 f_{cu}$ respectively. In all cases the amount of longitudinal reinforcement, $f_y = 250\,\text{N/mm}^2$, can be taken to be 0.2% of the gross cross-sectional area.

Example 5.6. Make an approximate initial design for a circular column required to withstand a design ultimate moment of 153 kN m and an axial load of 2400 kN for $f_{cu} = 50\,\text{N/mm}^2$ and $f_y = 425\,\text{N/mm}^2$.

Eccentricity of load $= 153/2400 = 0.0638$ m.

The size of the column is not yet known. Assume that the line of action of the axial load is inside the section, and check this later.

Cross-sectional area required $= 2400/(0.42 \times 50000) = 0.1143\,\text{m}^2$.

Diameter of column $= \sqrt{(0.1143/0.7854)} = 0.3815$ m, say 400 mm.

The line of action of the axial load is within the section. Total area of steel reinforcement $= 0.02 \times 0.1143 \times (425/250)\text{m}^2 = 3886\,\text{mm}^2$.

Example 5.7. Check the previous design using CP 110 Design Charts.

From *Table 3.2*, the steel would be eight 25 mm diameter bars. From CP 110, Table 19, suppose the cover to the links needs to be 25 mm. Again guided by CP 110, suppose that the links are of 8 mm diameter. Then the cover to the main steel is 33 mm. Referring to CP 110, Part 3, $h_s/h = (400 - 2 \times 33 - 25)/400 = 0.7725$. To be on the safe side use Chart 137 rather than

136. Now $M/h^3 = 0.153/0.4^3$ MN/m^2 = 2.39 N/mm^2 and $N/h^2 = 2.4/0.4^2$ = 15.0 N/mm^2, hence 100"A_{sc}"/A_c = 2.6. Therefore "A_{sc}" = 0.026 × 0.7854 × 400^2 = 3276 mm^2.

Eight reinforcement bars allow a system of rectangular stirrups, which are much better for construction purposes than helices. If the reinforcement is to be reduced it means choosing say either eight, six or twelve bars, for rectangular stirrups. No practical economy can therefore be made in the reinforcement (see *Table 3.2*) if the bars are all kept of the same diameter. Using two diameters, six 25 mm diameter and six 10 mm diameter bars can be used together in a symmetrical arrangement giving a steel area of 3416 mm^2 < 3927 mm^2 (eight 25 mm diameter bars). These will be positioned on a circle of diameter $h_s = 309$ mm. Thus spacing of bars will be π × 309/12 = 80.9 mm. If bars are alternate then distance between two consecutive bars = 80.9 − 12.5 − 5 = 63.4 mm. This is satisfactory if the size of coarse aggregate is less than 63–5 = 58 mm (see CP 110). It will probably be 25 mm and down aggregate because of steel from beams framing into this column, say. This column is similar to the one designed in CP 110, Part 3.

Example 5.8. Make an approximate initial design for a rectangular column required to withstand a design ultimate moment of 91 kN m and an axial load of 2460 kN for $f_{cu} = 50$ N/mm^2 and $f_y = 425$ N/mm^2. Then check the design using CP 110 Design Charts.

Eccentricity of load = 91/2460 = 0.037 m.

Assume that the line of action of the axial load is inside the section and check this later.

Cross-sectional area required = 2460/(0.45 × 50000) = 0.1093 m^2.

If one dimension is 450 mm, the other needs to be 0.1093/0.45 m = 243mm, say 250 mm. Thus the line of action of the axial load is within the section, as assumed.

Total area of steel reinforcement = 0.02 × 0.1093 × (250/425) m^2 = 1286 mm^2.

Use four 20 mm diameter bars. Using 30 mm cover to these bars, $d = 450 − 30 − 10 = 410$ mm, and $d/h = 410/450 = 0.91$. Use CP 110, Chart 88. Then $M/(bh^2) = 0.091/(0.25 × 0.45^2)$ MN/m^2 = 1.798 N/mm^2, $N/bh = 2.46/(0.25 × 0.45)$ MN/m^2 = 21.87 N/mm^2, and "A_{sc}" = 1.0 × 0.25 × 0.45/100 m^2 = 1125 mm^2. Hence the steel and size of section chosen are in order.

This column is similar to the one designed in CP 110, Part 2.

Chapter 6

Reinforced concrete frames and continuous beams and slabs

6.1 Bending moments and shear forces

CP 110 accepts frames being designed for bending moments and shear forces elastically. The second moments of areas are not usually varied according to the disposition of reinforcement. It is common practice to calculate the second moments of areas on the gross concrete cross sections only, ignoring reinforcement. The individual sections are then designed for ultimate limit states of bending moment and shear force. The disposition of this reinforcement influences the distribution of bending moments towards plastic collapse of a frame.

Much research has been done (e.g. the author has supervised the work of Refs. 1, 2, 3) with regard to the plastic redistribution of bending moments towards collapse. The fear is that if a designer chooses to make the resistance moment of a section excessively weak, then the section might fail by the extreme concrete fibre strain trying to exceed 0.0035 (the maximum amount experienced before concrete crushing), or in shear, before the other sections of the (structural steel type) collapse mechanism have realised their full resistance moments. To allow reasonable plastic redistribution of moments but to safeguard against the above, CP 110 says 'The ultimate resistance moment provided at any section of a member must not be less than 70% of the moment at that section obtained from an elastic maximum moments diagram covering all appropriate combinations of ultimate loads, and the elastic moment at any section in a member due to a particular combination of ultimate loads should not be reduced by more than 30% of the numerically largest moment given anywhere by the elastic maximum moments diagram for that particular member, covering all appropriate combinations of ultimate loads'. If this is done, however, CP 110 is concerned that the sections should be reasonably under-reinforced (because of the fear of concrete compression failure, which occurs suddenly). It therefore restricts neutral axis depth x to a minimum value according to the amount of the moment redistribution (see CP 110, Clause 3.2.2.3(4)). Also, for buildings of more than four storeys CP 110 more cautiously allows elastic moments to be reduced by only 10%, not 30% as mentioned previously.

Design in accordance with the first paragraph of this chapter is commendable, in that it automatically gives good control of crack widths and deflections (limit states of serviceability), because frames tend to behave elastically at working loads.

6.2 Continuous beams and slabs

The previous section deals with frames, but applies equally to continuous beams and slabs. *Table 6.1* gives bending moment coefficients for continuous beams or slabs whose spans are equal, or do not vary by more than say 10%, carrying uniformly distributed loads. For

Table 6.1.

Dead load	Live load
0.125 / 0.071 0.071	0.125 / 0.096 0.096
0.100 0.100 / 0.080 0.025 0.080	0.117 0.117 / 0.101 0.075 0.101
0.107 0.072 0.107 / 0.077 0.036 0.036 0.077	0.121 0.107 0.121 / 0.099 0.081 0.081 0.099
0.105 0.080 0.080 0.105 / 0.078 0.033 0.046 0.033 0.078	0.120 0.111 0.111 0.120 / 0.100 0.080 0.086 0.080 0.100

Table 6.2.

Dead load	Live load
0.38 0.62 / 0.62 0.38	0.44 0.62 / 0.62 0.44
0.40 0.50 0.60 / 0.60 0.50 0.40	0.45 0.58 0.62 / 0.62 0.58 0.45
0.39 0.54 0.46 0.61 / 0.61 0.46 0.54 0.39	0.45 0.60 0.57 0.62 / 0.62 0.57 0.60 0.45
0.40 0.53 0.50 0.47 0.60 / 0.60 0.47 0.50 0.53 0.40	0.45 0.60 0.59 0.58 0.62 / 0.62 0.58 0.59 0.60 0.45

live loads the coefficients are for complete spans loaded in the worst possible arrangement (see CP 110). The elastic bending moment at either support or span = Coefficient × Total load on span × Span. Similarly *Table 6.2* gives coefficients for shear forces. The elastic shear force at a support = Coefficient × Total load on span.

REFERENCES

1. Wilby, C. B., and Pandit, T., 'Inelastic behaviour of reinforced concrete single-bay portal frames', *Civil Eng. and Pub. Wks Rev.*, Mar. (1967)
2. Chapman, B. C., *Flexural Behaviour of Redundant Reinforced Concrete Frames*, PhD Thesis, University of Bradford (1973)
3. Noor, F. A., *Elastic and Inelastic Behaviour of Reinforced Concrete Frames*, PhD Thesis, University of Bradford (1970)

Chapter 7

Design using CP 110

7.1 Floor of building

Figure 7.1 shows a layout of columns, which has been determined to be sympathetic to the arrangement of the windows and layout of internal requirements (e.g. partition walls, equipment, machinery,

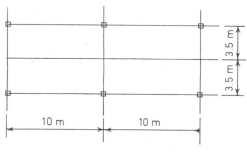

Fig. 7.1

etc.). The building is four 10 m bays wide and ten 7 m bays long. *Table 7.1* gives a very approximate guide for preliminary design proportioning. If there were no intermediate beams and the floor slabs were designed as 10 × 7 m two-way spanning they would be, from *Table 7.1*, about 10/40 m = 250 mm thick. This is a very thick slab. Intermediate beams reduce it considerably so that the total amount of concrete and reinforcement is less, and the load on the supporting beams, columns and foundations is less. Also, the shuttering does not need to be as strong. The intermediate beams can be as shown in *Figure 7.1* and from *Table 7.1* the slab is about 3.5/35 m = 100 mm, say 150 mm thick, as this is about a minimum floor thickness for practical reasons, and for deflection in this example, see later. Otherwise they could have been at right-angles to these, giving two-way spanning slabs 7 × 5 m of thickness, from *Table 7.1* approximately 7/40 m = 175 mm. If these two schemes are compared, the

124

first is favoured as the shorter beams carry a greater proportion of the load on each 10 × 7 m panel.

7.1.1 Floor slab

This is therefore a 150 mm thick one-way slab, continuous for 20 bays, each of 3.5 m span. Suppose the floor carries bedrooms for a hotel or hospital. CP 3 requires the floor to be designed for a uniformly distributed load of 2 kN/m². As the slab was made thicker than required for practical reasons the concrete will only need to be weak, but as it is a slab and not very thick we do not want the cover to be very great. Considering mild exposure on CP 110, Table 19, we choose Grade 25 concrete so that we can have 20 mm cover—not 25 mm as with Grade 20. Referring to CP 110, Table 56, the slab can have $1\frac{1}{2}$ h fire resistance and we assume that this is satisfactory. The floor will carry lightweight partitions and there will be floor finishes, perhaps tiles on the floor, and either plaster or a suspended ceiling and minor services below; assume all this weighs 1.5 kN/m². The self-weight of the floor (taking the weight density of reinforced concrete as 23.6 kN/m³) = 0.15 × 23.6 = 3.54 kN/m². Thus characteristic dead load = 3.54 + 1.5 = 5.04 kN/m². The building is wide and long enough compared to its height for wind forces to be neglected (see CP 3). From CP 110, Clause 2.3.3.1, design load = 1.4 × 5.04 + 1.6 × 2 = 10.25 kN/m². From CP 110, Table 4, maximum bending moment is at the first interior support and = 10.25 × 3.5²/9 = 13.95 k N m per metre width of slab. Using f_y = 460 N/mm² and CP 110 Design Chart No. 4, and (guessing 12 mm diameter bars and thus d = 150 − 26 = 124 mm) $M/(bd^2)$ = 0.013 95/0.124² MN/m² = 0.9073 N/mm², then $100A_s/(bd)$ = 0.235, ∴ A_s = 0.235 × 0.124/100 = 0.000 291 4 m². From *Table 3.2* use 10 mm diameter bars at 250 mm centres. This is reasonable for detailing. The steel at other locations can be obtained *pro rata* to the bending moments of CP 110, Table 4. For example, the smallest of these is at the middle of the interior spans and as Chart No. 4 is linear for smaller values of M, $100A_s/bd^2$ = 0.235 × 9/14 = 0.1511. The value of d could be revised to 0.125 m.

To check that the deflection is not excessive, CP 110, Table 8, allows a span-to-effective-depth ratio = 26 and from CP 110, Table 10, the modification factor can be taken as 1.41. Therefore allowable maximum span = 1.41 × 26 × 0.125 = 4.58 m > 3.5.

With regard to the limit state of cracking, CP 110, Clause 3.11.8.2, says that for normal conditions of exposure no special check is required if our slab thickness is less than 200 mm thick, which it is;

and the clear distance between the reinforcement bars must not exceed 3*d*, which for the 10 mm diameter bars = 3 × 125 = 375 mm, and this can be followed in the detailing.

We need some steel at right-angles to the above steel. This is usually called *distribution steel*; it helps to distribute point loads across the width of a slab, to resist shrinkage and temperature stresses, and to help fix the main steel. Using high yield distribution steel reference to CP 110, Clause 3.11.4.2, gives the area of this steel = 0.0012 × 0.15 = 0.000 180 m^2 per metre.

Table 7.1.

Ratios of span to overall depth	
Simply supported beams	20
Continuous beams	25
Cantilever beams	10
Slabs spanning in one direction, simply supported	30
Slabs spanning in one direction, continuous	35
Slabs spanning in two directions, simply supported	35
Slabs spanning in two directions, continuous	40
Cantilever slabs	12

It is very unlikely that shear reinforcement will be required (in this eventuality we would normally avoid having to use it by making the slab thicker). From CP 110, Table 4, maximum shear force = 0.6 × 10.25 × 3.5 = 21.53 kN per metre width of slab. Referring to Section 3.4, V/bd = 21.53/0.125 kN/m^2 = 0.1722 N/mm^2, which is obviously satisfactory from CP 110, Table 5.

7.1.2 Beams of 10 m span

CP 110, Table 19, gives a minimum cover of 20 mm for mild exposure and Grade 25 concrete. Using a fire resistance of $1\frac{1}{2}$ h as for the slab, CP 110, Table 54, requires a minimum concrete cover to main reinforcement of 15 mm and requires a beam width of 85 mm, using vermiculite/gypsum plaster.

The continuous beam supporting the heaviest loading is the penultimate beam. From CP 110, Table 4, the reaction on this beam from the slab = (0.6 + 0.55) × Ultimate load. Hence dead load from slab = 1.15 × 5.04 × 3.5 = 20.29 kN/m and the live load from slab = 1.15 × 2 × 3.5 = 8.05 kN/m. From *Table 7.1* the overall depth of the beam required is approximately 10/25 = 0.4 m.

Within reason the greater the depth the more economic and easy for design, detailing and fixing of the reinforcement. A small amount of extra vertical shutter (which does not alter scaffolding costs) and of concrete can save expensive reinforcement and its fixing, and reduce concreting costs of placing concrete around high percentages of reinforcement. Architects often require the overall depths of beams to be a reasonable minimum for reasons of aesthetics. Deeper beams increase the heights of buildings where strict use is made of minimum headrooms, but we are only talking about altering the depth of beams by perhaps about 0.1 m or so to increase the economy and speed of the reinforced concrete construction. In this example suppose the architect for aesthetic reasons does not desire a beam with overall depth more than 0.4 m. The breadth of the rib of a beam will often be about 1/3 to 1/2 of the overall depth, with a minimum sufficient to accommodate three 25 mm diameter bars. Using 19 mm down coarse aggregate the horizontal distance between bars, from CP110, must be greater than $19 + 5 = 24$ mm, say 25 mm. Hence width of rib to accommodate three 25 mm diameter bars $= 5 \times 25 + 2 \times 25$ (i.e. covers) $= 175$ mm. Hence use a beam of overall depth 0.4 m and breadth of rib of 0.2 m. Then self-weight of rib $= (0.4 - 0.15) \times 0.2 \times 23.6 = 1.18$ kN/m. Hence for design purposes if a span is unloaded (CP110, Clause 3.2.2) it carries its minimum dead load of $20.29 + 1.18 = 21.47$ kN/m. If loaded it carries a design total ultimate load of $1.4 \times 21.47 + 1.6 \times 8.05 = 42.94$ kN/m. The first part of CP110, Clause 3.2.2, means that all spans always support the minimum self-weight of 21.47 kN/m, but the excess of $42.94 - 21.47 = 21.47$ kN/m can be disposed on various spans. *Tables 6.1* and *6.2* can be used if the minimum self-weight is considered as dead load and the excess of total ultimate load over minimum self-weight is considered as live load (by chance dead and live load are equal in this example). Hence maximum bending moment is at penultimate support $= (0.107 + 0.121) \times 21.47 \times 100 = 489.5$ kN m and maximum shear force is adjacent to inner support of either end span $= (0.61 + 0.62) \times 21.47 \times 10 = 264.1$ kN. At this support the beam is cracked in flexure in the top and acts as a rectangular beam. The overall depth of the beam is 0.4 m. The slab has top main steel up to 10 mm diameter with 20 mm cover. The main beam steel over the support must be beneath this slab steel, hence assuming it is of 25 mm diameter bars its effective depth $= 400 - 30 - 12.5 = 358$ mm. If we decide to use reinforcement with $f_y = 425$ N/mm^2, then using CP110, Part 2, Design Chart No. 3, $M/(bd^2) = 489.5/(0.2 \times 0.358^2)$ kN/m$^2 = 19.1$ N/mm^2.

If we decide to use reinforcement with $f_y = 425$ N/mm^2, then an examination of CP110, Part 2, Design Charts Nos. 3 and 33, shows that the beam is much too small. Even a much stronger concrete (see Chart No. 40) is inadequate, although it would normally be undesirable to have a different concrete mix in the rib to the slab, and undesirable to make the slab of a stronger mix than required to suit the rib. V/bd = 264.1/(0.2 × 0.358) kN/m^2 = 3.69 N/mm^2. From CP 110, Tables 5 and 6, this is too great. Detailing for shear is often difficult, so if we keep V/bd down to about half of the maximum value of Table 6, namely 3.75/2 = 1.88 say, then $d \simeq$ 264/(0.2 × 1880) = 0.702 m. Then overall depth $\simeq$ 702 + 13 + 30 = 745 mm, which is rather deep, and *Table 7.1* has not guided us too well. This is because the spans are rather long for this loading for reinforced concrete.

Hence, let us revise the design by dividing each 7 m distance between columns (see *Figure 7.1*) into three instead of two, with beams of 10 m span. This means that the slab design of Section 7.1.1 needs revision. Its thickness would perhaps be reduced to 125 mm, but to reduce its weight further we will use a 125 mm thick hollow-tile floor weighing 2.14 kN/m^2. Hence, characteristic dead load of slab = 2.14 + 1.5 = 3.64 kN/m^2. Using the size of beam obtained previously from *Table 7.1*, the minimum dead load is 1.15 × 3.64 × 7/3 + 1.30 = 11.07 kN/m. The design total ultimate load is 1.4 × 11.07 + 1.6 × 1.15 × 2 × 7/3 = 24.08 kN/m. The maximum bending moment at a penultimate support = 0.107 × 11.07 × 100 + 0.121 × (24.08 − 11.07) × 100 = 275.9 kN m and maximum shear force is 0.61 × 11.07 × 10 + 0.62 × (24.08 − 11.07) × 10 = 148.2 kN. Now $M/(bd^2)$ = 275.9/(0.2 × 0.358^2) kN/m^2 = 10.76 N/mm^2. Chart 33 still shows this to be undesirably high. At mid span the beam is a T-beam and the bending moment might be slightly less. Hence it is desirable in this instance to reduce the bending moments at the supports, assuming plastic action and moment redistribution, as this will reduce the compression steel required at the supports perhaps without requiring such steel at mid spans. Assuming the structure is three storeys in height, then CP 110, Clause 3.2.2.3, allows this bending moment to be reduced by 30% if $x \leqslant 0.3d$. Thus $M/(bd^2)$ = 0.7 × 10.76 = 7.53 N/mm^2. Then from Chart 33 A_s = 2.35 × 200 × 358/100 = 1683 mm^2 and A'_s = 2.0 × 200 × 358/100 = 1432 mm^2. From *Table 3.2* use three 25 mm diameter bars and one 20 mm diameter bar as tension steel and three 25 mm diameter bars as compression steel. The previously estimated width of the rib allowed for this amount of steel. In the top over the support the 20 mm diameter bar will be in a layer beneath the 25 mm diameter bars. Allowing a gap of 19 × 2/3 = 13 (i.e. 2/3 of maximum aggregate size) between the adjacent 20 mm and 25 mm bars; then d = (1473 × 358 + 314 × 322)/1787 = 352, which is < 358. Hence check its correctness. $M/(bd^2) = 0.7 \times 275.9/(0.2 \times 0.352^2)$ kN/m^2 = 7.79 N/mm^2. Then from Chart 33, $A_s = 2.46 \times 200 \times 352/100 = 1732$ mm^2 and A_s' is unaltered. The A_s provided is 1787 mm^2 and thus satisfactory. $V/bd = 148.2/(0.2 \times 0.352)$ kN/m^2 = 2.105 N/mm^2 < 3.75 of CP 110, Table 6. Then see Example 3.8. At support $100A_s/bd = 100 \times 1787/(200 \times 352) = 3.54$. From CP 110, Table 5, shear resistance provided by concrete alone = $(0.85 + 0.54 \times 0.05) \times 200 \times 352$ N = 61.7 kN. Hence shear reinforcement is required and has to resist $148.2 - 61.7 = 86.5$ kN. Using stirrups the V/d required is $86.5/0.352 = 246$ N/mm. From *Table 3.5* use 8 mm diameter four-arm stirrups at 175 mm centres, using steel with $f_{yv} = 250$ N/mm^2.

There are significant bending moments and shear forces because of the ends not being pin-jointed to the external columns. This reduces the maximum shear forces and bending moments used previously, and hence this beam can be designed and detailed with regard to ultimate limit state from the foregoing.

Considering the maximum bending moment within an end span, this being greater than for internal spans, using *Table 6.1* this is $0.077 \times 11.07 \times 100 + 0.099 \times 13.01 \times 100 \times 214.0$ kN m. But because the bending moment at the penultimate support was reduced by about $0.3 \times 275.9 = 82.8$ kN m the moment of 214.0 must be increased by $0.5 \times 82.8 = 41.4$, to give 255.4 kN m. Taking the centre of the longitudinal compressive force as acting at half the depth of the flange of this T-beam, $z \simeq 358 - 63 = 295$ mm. (Note that the cover could be reduced at mid span, increasing z slightly.) The design stress of the reinforcement, using $\gamma_m = 1.15$, $= 425/1.15 = 369.6$ N/mm^2. Hence the area of tension steel required = $255.4 \times 10^6/(369.6 \times 295) = 2342$ mm^2. From *Table 3.2* use a bottom layer of reinforcement of three 25 mm diameter bars and a second layer of three 20 mm diameter bars. Allowing a gap of 13 mm between the layers it is obvious that d is less than assumed. Some of the 20 mm bars may have to be increased to 25 mm, hence assume $d = 400 - 20 - 8 - 25 - 13/2 = 340$ mm. Then $z = 340 - 63 = 277$ mm, and $A_s = 255.4 \times 10^6/(369.6 \times 277) = 2495$ mm^2. From *Table 3.2* use a bottom layer of three 25 mm diameter bars and a second layer of two 25 mm diameter and one 12 mm diameter bars. Breadth of flange required is obtained (see Section 3.7.5) from $255.4 \simeq 0.4 \times 25\,000 \times b \times 0.125 \times 0.277$, $\therefore b = 0.738$ m. The hollow tiles can be kept out of this breadth of slab at each beam position. A more accurate design of this T-section can be effected as in Example 3.18.

With regard to the limit state of deflection, the worst span-to-effective-depth ratio $= 10/0.34 = 29.41$, $A_s = 2455$ (five 25 mm) + 113 (one 12 mm) = 2568 mm^2, and $100A_s/(bd) = 0.2568/(0.2 \times$

0.340) = 3.78. From CP 110, Table 10, modification factor = 0.76, so using CP 110, Table 8, the allowable span-to-effective-depth ratio = 0.76 × 26 = 19.76, which becomes 0.8 × 19.76 = 15.81, as the beam is a T-beam with web width to effective flange width = 0.2/0.738 = 0.271 (see CP 110, Clause 3.3.8.2). The beam is thus unsatisfactory in this respect and to endeavour to rectify this situation the depth can be increased. This will also have the effect of reducing the reinforcement. Suppose the effective depth is increased to 10/15.81 = 0.633, say 0.66 m. Then the A_s will be slightly more than 2495 × 277/(660 − 63) = 1158 mm^2 because the weight of the beam is increased by 0.2 × (0.66 − 0.340) × 23.6 = 1.510 kN/m. Total ultimate design load is increased from 24.08 kN/m by 1.4 × 1.510 = 2.114 kN/m, i.e. by 8.8 %. Hence $A_s \simeq 1.088 \times 1158 = 1260$ mm^2. Then $100A_s/bd \simeq 0.1260/(0.2 \times 0.66) = 0.955$, and from Table 10, modification factor = 0.98 + 0.08 × 0.045/0.25 = 0.994. Then using Table 8 and Clause 3.3.8.2, the allowable span-to-effective-depth ratio = 0.8 × 0.994 × 26 = 20.67. Hence the limit state of deflection is now satisfactory, but the previous design needs revision. The limit state of cracking is easy to comply with in the detailing (see CP 110, Clause 3.3.9 and Table 24).

As mentioned before, bending moments due to the beams framing into the external columns can be distributed along the continuous beams advantageously if a more laborious design is to be made. In any case the bending moment in the beam at this junction must be assessed as given at the end of Section 7.1.3 for detailing the beam at this location.

7.1.3 External columns between ground and first floor

Figure 7.2 shows an external column. The base shown rests on a cohesive soil (clay) and is designed for uniform soil pressure. That is, the base is assumed to rotate because of the inelastic or plastic action (or creep) of the soil. So that the shutters can be unaltered for economy the external column is designed for BC and the higher portions kept the same size, the steel being reduced. The greatest vertical load is little greater at B than at C, yet it is combined with a substantial bending moment at C, which is the critical section for design.

Considering durability (mild exposure) and fire resistance ($1\frac{1}{2}$ h) CP 110, Tables 19 and 59, mean that the cover of Grade 25 concrete to the links needs to be 20 mm and the minimum dimension of the concrete needs to be 150 mm, using vermiculite/gypsum plaster.

The vertical loads can be accurately obtained from the shear forces

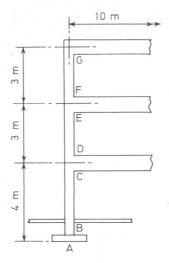

Fig. 7.2

of the beams framing into the columns and estimating the self-weight of the columns. The 7 m long beams should therefore be designed similarly to Section 7.1.2 before the columns are designed. Assume that the vertical load at C comprises a characteristic dead load of 665 kN and a maximum characteristic live load of 166 kN. To estimate the size of the column assume it is axially loaded, ignore the strength of its reinforcement, and increase the cross-sectional area by about 50%. Then the design ultimate load = $1.4 \times 665 + 1.6 \times 166 = 1197$ kN. Then cross-sectional area of column required (see equation 5.2) = $1.5 \times 1\,197\,000/(0.4 \times 25) = 179\,500$ mm^2, say 400 × 450 mm. Second moment of area for the column = $400 \times 450^3/12 = 3040 \times 10^6$ mm^4. The stiffnesses of columns DE and CB are $3040 \times 10^6/3000 = 1\,013\,000$ mm^3 and $3040 \times 10^6/4000 = 759\,000$ mm^3 respectively. The second moment of area of the beam poses a problem, as the beam is a T-beam at mid span but a rectangular beam in effect at the location of cracks near the supports. The writer considers the former as the more accurate assumption, as did Scott and Glanville, and many structures have been designed on this basis. However, the latter assumption is easier for calculation, gives higher moments in the columns, and is favoured by Allen (1974), Higgins and Hollington (1973) of the Cement and Concrete Association. Using the latter assumption (depth of beam, say = $660 + 60 = 720$ mm, say 725 mm) the second moment of area for the beam =

$200 \times 725^3/12 = 6351 \times 10^6$ mm^4 and the stiffness $= 6351 \times 10^6/10\,000 = 635\,100$ mm^3. For this beam total design ultimate load $= 24.08 \times 1.088 = 26.2$ kN/m, and fixed end moment $= 26.2 \times 10^2/12 = 218.3$ kN m. The bending moment at C, from CP110, Clause 3.2.2.1, is $218.3 \times 759/(759 + 1013 + 635/2) = 79.3$ kN m. As A in *Figure 7.2* is assumed to be in effect a hinge, it would be more accurate to reduce the stiffness of CB, but CP110 does not suggest this, and gives a higher moment at C.

There are walls between the external columns, various internal walls, and the overall height of the building compared to its horizontal dimensions was sufficiently low for lateral wind forces to be ignored. It can be assumed therefore that the beam column junctions will not move laterally, i.e. the columns can be considered as *braced* as defined by CP110, Clause 3.5.1.3. From CP110, Table 15, take the effective height of column CB as the length CB $= 3.7$ m, guessing AB as 0.3 m. Then $3.7/0.4 = 9.25 < 12$, hence column can be treated as a short column (see Section 5.2).

From CP110, Clause 3.5.5, the minimum design ultimate bending moment $= 1197 \times 0.05 \times 0.45 = 26.9$ kN m < 79.3 kN m, hence design for the latter. If the links are 8 mm diameter, this means that the cover to the main steel is 28 mm, say 30 mm. Suppose 25 mm bars are to be used in a single layer at each side of the column, then effective depth $= 450 - 43 = 407$ mm and $d/h = 407/450 = 0.90$. Thus for $f_{cu} = 25$ N/mm^2 and $f_y = 425$ N/mm^2, use Design Chart 76 of CP110, Part 2. $N/bh = 1.197/(0.4 \times 0.45)$ MN/m$^2 = 6.65$ N/mm^2, $M/(bh^2) = 0.0793/(0.4 \times 0.45^2)$ MN/m$^2 = 0.979$ N/mm^2, therefore $A_{sc} = 0.00\%$ of $b \times h$. But this percentage should not be less than one (see CP110, Clause 3.11.4.1) hence a smaller column could be designed or 1 % of $b \times h$ can be used as reinforcement. In this case not only would this column CB be uneconomic but also columns DE and FG, as the shutters were to be kept the same for economy. The design of the column should therefore be revised.

With the present design the bending moment in the beam framing into the column at its support is the sum of the bending moments in the columns at C and D, i.e. equal to $218.3\,(759 + 1013)/(759 + 1013 + 635/2) = 185$ kN m. This is undesirably high for the beam (see Section 7.1.2) and a re-design of the columns, making them more slender, will reduce this bending moment. The beam steel framing into the column would be excessive for detailing.

7.1.4 Bases

Suppose the column CB is re-designed so that the vertical design ultimate load at B is 1215 kN. We guessed previously that the thickness of the base is 0.3 m, and this gives a characteristic pressure on the soil of $0.3 \times 23.6 = 7.08$ kN/m^2 and an ultimate design pressure of $1.4 \times 7.08 = 9.91$ kN/m^2. As mentioned before the soil is cohesive and is considered to give a uniform pressure beneath the base. Assume the soil beneath the base can safely withstand a pressure of 217 kN/m^2. Using a load factor of say 1.8, the ultimate pressure on the soil can be $217 \times 1.8 = 390$ kN/m^2. Then the area of the base needs to be $1215/(390 - 9.91) = 3.2$ m^2. Making it square to save shuttering, it needs to be 1.8×1.8 m.

From CP 110, Table 19, cover to reinforcement for moderate exposure (buried concrete) = 40 mm. If 20 mm bars are to be used to form a square mesh, then the effective depth for the bars in the upper layer.= $300 - 40 - 30 = 230$ mm. Suppose on re-design the column became 350×410 mm.

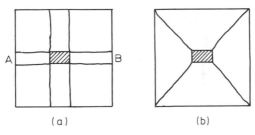

(a)   (b)

Fig. 7.3

For shear using CP 110, Clause 3.10.4.2, condition (1): ultimate design uniform pressure on base = $1215/1.8^2 = 375$ kN/m^2, shear force on section distance 1.5 times effective depth = $1.5 \times 230 = 345$ mm from face of column = $375 \times 1.8 \times (1.8/2 - 0.35/2 - 0.345) = 256.5$ kN, $V/bd = 0.2565/(1.8 \times 0.23)$ MN/m^2 = 0.62 N/mm^2. Condition (2): critical shear perimeter is $1.5 \times$ Slab thickness = $1.5 \times 300 = 450$ mm away from the faces of the column and is $= 2 \times 350 + 2 \times 410 + 2\pi \times 450 = 4347$ mm, area enclosed by this perimeter = $350 \times 410 + 2 \times 450 \times 350 + 2 \times 450 \times 410 + \pi \times 450^2 = 1\,464\,000$ mm^2. Shear on perimeter = $1215 - 1.464 \times 375 = 666$ kN, shear stress on perimeter = $0.666/(4.347 \times 0.23) = 0.666$ N/mm^2. From CP 110, Tables 5, 6 and 14, these shear stresses will be all right if the longitudinal reinforcement is just more than 1 % of $b \times d$.

With regard to ultimate bending moment in slab, by yield line theory the slab can fail as in *Figure 7.3(a)* or (b). The former requires an ultimate moment per unit length = $m = 1215/24 = 50.6$ kN m/m,

K

whereas the latter (considering the yield line AB) requires 1.8 × 0.5 × $[0.5(1.8 - 0.35)]^2$ × 375/1.8 = 98.6 kN m/m. We must design for the latter, which is as recommended by CP110, although CP110 does not give the basis of its recommendation. For $f_{cu} = 25$ N/mm^2 and $f_y = 425$ N/mm^2 use Design Chart 3 of CP110, Part 2, $M/(bd^2)$ = 0.0986/0.23^2 MN/m^2 = 1.864 N/mm^2, ∴ $A_s = 0.56\%$ of bd = 0.0056 × 1000 × 230 = 1288 mm^2/m; from *Table 3.2* use say 20 mm diameter bars at 225 mm centres, giving $A_s = 1396 = 0.61\%$ of bd. This is inadequate for the shear stresses of the last paragraph (Table 5). Either the A_s can be increased accordingly or the design revised with a thicker slab. The latter is favoured as it helps with detailing the compression anchorage of the vertical column bars into the bases. Sometimes just strengthening the concrete would solve the inadequacy but from Table 5 even with Grade 40 concrete $A_s = 0.61\%$ of bd only gives an ultimate shear stress of 0.59 N/mm^2, which is inadequate in this instance. Extra concrete thickness of a base is usually very inexpensive because rough or no shutters are used for the edges and concreting at this level is very inexpensive, involving no scaffolding, etc., or special finishing.

Increasing the slab thickness of a base means that the total load on the soil will be greater and thus the base will have to be larger, increasing the maximum bending moment, but the longitudinal steel in this instance is not great. Guided by Table 5, if the maximum ultimate shear stress of the design is reduced to about 0.5 N/mm^2, then $d \simeq 230 \times 0.666/0.5 = 306$ mm. Slab thickness then = 306 + 70 = 376 mm, say 400 mm, and $d = 400 - 70 = 330$ mm. This increased depth is best made keeping the soffit of the base at the same level. The extra design pressure due to the base is 1.4 × 0.1 × 23.6 = 3.3 kN. The area of the base then needs to be 1215/(390 − 9.91 − 3.3) = 3.22 m^2 so the 1.8 × 1.8 m base is still satisfactory. For shear using Clause 3.10.4.2, condition (1): shear force on section distance 1.5 times effective depth = 1.5 × 330 = 495 mm from face to column = 375 × 1.8 × (0.9 − 0.175 − 0.495) = 155.3 kN, V/bd = 0.1553/(1.8 × 0.33) = 0.261 N/mm^2. Condition (2): critical shear perimeter is 1.5 × 400 = 600 mm away from the faces of the column and = 2 × 350 + 2 × 410 + 2π × 600 = 5290 mm, area enclosed by this = 350 × 410 + 2 × 600 × 350 + 2 × 600 × 410 + π × 600^2 = 2186 000 mm^2, shear on perimeter = 1215 − 2.186 × 375 = 395 kN, shear stress on perimeter = 0.395/(5.29 × 0.33) = 0.226 N/mm^2. The ultimate bending moment required per unit length is as before, then using Design Chart 3, $M/(bd^2) = 0.0986/0.33^2 = 0.905$ N/mm^2, ∴ $A_s = 0.252\%$ of $bd = 832$ mm^2/m; from *Table 3.2* use 16 mm diameter bars at 225 mm centres. From Table 5 the above shear stresses are now satisfactory.

With regard to limit state of cracking the distance between bars can be $180 \times 2 = 360$ mm from CP 110, Clause 3.11.8.2, so the 225 mm suggested is correct. Checking local bond: condition (2) perimeter of steel $= \pi \times 16/0.225 = 223$ mm/m. From CP 110, equation 43, the ultimate local bond stress $= V/(d\Sigma u_s) = 0.395/(0.33 \times 0.223 \times 5.29) = 1.015$ N/mm^2; condition (1) ultimate local bond stress $= 0.1553/(0.33 \times 0.223 \times 1.8) = 1.172$ N/mm^2. From CP 110, Table 21, both are less than 2.5 N/mm^2 (deformed bars) and therefore satisfactory.

The reinforcement must have adequate anchorage length within the size of the base. If the bars are to be without end anchorages, using *Table 2.5* $l_b = 56d_b = 56 \times 16 = 896$ mm. The distance available $= 0.9 - 0.35/2 - 0.04$ (end cover) $= 0.685$ m, which is inadequate. Hence the diameter of the bars is limited to $685/56 = 12.2$, say 12 mm. Thus the 16 mm bars should from *Table 3.2* be replaced by 12 mm diameter bars at 125 mm centres.

7.1.5 Anchorage of column bars into bases (*see Sections 2.6–2.6.10*)

In this example the column bars are in compression. There will be *starter bars* projecting from each base, as shown in *Figure 7.4*. These

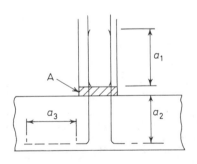

Fig. 7.4

are lapped with a *compression lap* with the column bars. Distance a_1 is this lap plus a tolerance of say 20 mm. A is a *kicker pad* of concrete, say 50 mm deep, for holding the column shutters apart at this point. The base must be adequately thick to accommodate the distance a_2, which needs to be the *compression anchorage length*.

Extra length such as a_3 cannot be counted in the compression lap for similar reasons to those given in Section 2.6.9. If the base is too thin to accommodate a_2 then it may need to be thickened, giving economies in the reinforcement for bending moments in the base. Alternatively, larger diameter or more starter bars, or both, may be used.

7.1.6 Student design office exercise

Each member of the class can be given a different column grid layout similar to that of *Figure 7.1*, i.e. 10×7 m, $9.9 \times (0.7 \times 9.9)$, $9.8 \times (0.7 \times 9.8)$ etc. A typical student can check calculations at all stages, with his colleagues working on grids immediately on either side of his own. This helps supervision enormously.

The exercise can be as in Sections 7.1–7.1.3, can be revised as required to more economic designs, and can be more accurately designed using bending moment envelopes.

Pairs of students can design structures of the same geometry, one with Grade 25 concrete and the other with Grade 20 or Grade 30 concrete. Pairs of students can also design structures of the same geometry and grade of concrete but using different steels.

7.1.7 Floor of building

Suppose in this scheme that the floor of Section 7.1 is supported by a 5 m square, as opposed to the rectangular system, of columns. It seems natural, because of symmetry, to choose two-way spanning slabs or a flat slab, rather than a system of one-way spanning slabs with subsidiary and main beams. Using *Table 7.1* the two-way continuous slabs would need to be approximately $5000/40 = 125$ mm thick. This is reasonably thin, hence an intermediate system of crucifix beams, making the slabs 2.5 m square, is not required. The beams between the columns supporting the slab, from *Table 7.1*, will need an overall depth of about $5000/25 = 200$ mm, and breadth say about half of this, namely 100 m, say 125 mm as 100 mm is rather too small to accommodate beam reinforcement.

Ignoring shear, a flat slab needs an overall depth of about $125/0.9 = 139$ mm, say 150 mm. With drops the slab would need to be about 125 mm thick and the drops about $1.4 \times 125 = 175$ mm thick. In both cases it would be normal to avoid the need for shear reinforcement and this would usually necessitate the slab being thicker even if column heads are used.

7.1.8 Design tables

Table 7.2 is useful for the design of the beams shown and also for giving fixed end moments for commencing moment distribution analyses. In this table, as regards the end restraints, F denotes free to rotate and C denotes constrained (i.e. fixed or encastré). The bending moments at A, B and C respectively are $\alpha_A Ql$, $\alpha_B Ql$, $\alpha_C Ql$ respectively, where Q is the total load on span l, and C is the position of maximum

Table 7.2.

Loading	End restraint A	End restraint B	α_A	α_C	α_B	Deflection β	P γ
	F	F	—	1/8	—	1/76.8	1/2
	C	F	$-1/8$	1/14.2	—	1/185	0.625
	C	C	$-1/12$	1/24	$-1/12$	1/384	1/2
	F	F	—	1/4	—	1/48	1/2
	C	F	$-1/5.33$	1/6.4	—	1/107.3	0.688
	C	C	$-1/8$	1/8	$-1/8$	1/192	1/2
	F	F	—	1/6	—	1/60	1/2
	C	F	$-1/6.4$	1/9.51	—	1/139.5	0.656
	C	C	$-1/9.6$	1/16	$-1/9.6$	1/274.3	1/2
	F	F	—	$\alpha\alpha_1$	—	$\alpha^2\alpha_1^2/3$	α_1
	C	F	$\alpha\alpha_1\alpha_2/2$	$\alpha^2\alpha_1\alpha_3/2$	—		$\alpha_1\alpha^1$
	C	C	$-\alpha\alpha_1^2$	$2\alpha^2\alpha_1^2$	$-\alpha^2\alpha_1$		$\alpha_1^2(1 + 2\alpha)$

Table 7.3.

No. of spans	Bending moment coefficients						Shear force coefficients									
	A	B	C	D	E	F	AB	BA	BC	CB	CD	DC	DE	ED	EF	FE
2	−1.00	+0.25	—	—	+0.25	—	+1.250	−1.250	—	—	—	—	—	—	−0.250	+0.250
3	−1.00	+0.267	—	—	−0.067	—	+1.267	−1.267	−0.333	+0.333	—	—	—	—	+0.067	−0.067
4	−1.00	+0.268	−0.071	—	+0.018	—	+1.268	−1.268	−0.339	+0.339	+0.089	−0.089	—	—	−0.018	+0.018
5	−1.00	+0.268	−0.072	+0.019	−0.005	—	+1.268	−1.268	−0.340	+0.340	+0.091	−0.091	−0.024	+0.024	+0.005	−0.005
2	−1.00	+0.500	—	—	+0.500	−1.00	+1.500	−1.500	—	—	—	—	—	—	−1.500	+1.500
3	−1.00	+0.200	—	—	+0.200	−1.00	+1.200	−1.200	0	0	—	—	—	—	−1.200	+1.200
4	−1.00	+0.286	−0.143	—	+0.286	−1.00	+1.286	−1.286	−0.429	+0.429	+0.429	−0.429	—	—	−1.286	+1.286
5	−1.00	+0.263	−0.053	−0.053	+0.263	−1.00	+1.263	−1.263	−0.316	+0.316	0	0	+0.316	−0.316	−1.263	+1.263

Table 7.4 Weights of materials

	kN/m^3		kN/m^2
Aluminium	27.0	Concrete hollow tile slabs	
Ashes (dry)	6.3	125 mm thick	2.14
Asphalt	20.4	150 mm thick	2.38
Brickwork, cement mortar		190 mm thick	2.68
common brick	19	Corrugated sheeting	
pressed brick	23	galvanised iron	0.144
Cement		asbestos-cement	0.156
loose	11.8–13.3	Doors	0.384
bags	11.0–12.6	N-light roof glazing	0.264
bulk	12.6–14.1	Roofing felt (two-layer built up)	0.048
Coal		Windows	0.240
solid	12.8		
crushed washed	9.0		
crushed unwashed	9.3		
Concrete			
plain or reinforced	23.6		
granolithic or terrazzo	23.6		
foamed slag non-structural	13–15		
foamed slag structural	21		
aerated	8.5–9.4		
Cork	2.4		
Copper	85.9		
Fibreboard	2.9		
Fibreboard, compressed	5.0		
Glass	24–27		
Iron	70.6		
Lead	112		
Lime plaster	18.8		
Macadam	21		
Mortar (set)			
cement screeds	22.6		
lime screeds	15.7–17.3		
Plasterboard	9.3		
Rubber	9.6		
Steel (cast or mild)	77		
Tarmacadam	23		
Vermiculite/cement screed	5.8		
Wood paving	8.7		
Wood wool/cement slabs	5.8–7.2		
Woodwork			
red pine	4.8–7.2		
teak	6.4–8.8		
pitch pine	6.6–7.2		
greenheart	10–12		

positive bending moment in the span. The maximum deflection along the span is $\beta Q l^3$. The reaction at A is $P = \gamma Q$. Also $\alpha_1 = 1 - \alpha$, $\alpha_2 = 2 - \alpha$, $\alpha_3 = 3 - \alpha$ and $\alpha^1 = 1 + \alpha - \alpha^2/2$.

Table 7.3 is very useful in conjunction with *Tables 7.2, 6.1* and *6.2*. It is for continuous beams of spans AB, BC, CD, etc. The first section is for a unit bending moment applied at A, whilst the second section is for unit bending moments applied simultaneously at A and the other end of the continuous beam. The bending moment at any support is the applied bending moment M at the end (or ends) times the coefficient. The shear force next to any support is $M \times$ Shear force coefficient divided by the span. EF is always the end span, otherwise the spans read consecutively from left to right (i.e. AB, BC, CD, etc.).

Table 7.4 gives the weights (for $g = 9.807$ m/s^2) of various building materials.

Chapter 8

Prestressed reinforced concrete

8.1 Prestressing

Prestressing consists of initially applying loads to a member to counteract the effects of the working loads to which it will eventually be subjected. Concrete is relatively weak in tension compared with compression, so the prestressing forces are used to compress zones which will subsequently be required to carry tension. Prestressing forces are usually applied in one of the following ways:

1. Stretching wires, cables or bars on a bed, concreting the member around such wires, and then releasing the wires when the concrete is sufficiently hard. When the wires are released, they shorten, and compress the concrete member, the line of action of such compression for each wire being the profile of the wire in the beam. This procedure is known as *pretensioning*.

2. A member is concreted and a duct is formed in the member either with an inflatable tube or with a metal sheath. A tendon, consisting of either a bar (e.g. Lee McCall bar), cable (e.g. Strand) or groups of wires (e.g. Freyssinet and Gifford-Udall systems), is threaded through the duct and tensioned when the concrete is sufficiently hard, and anchored to the concrete member, so that the concrete member is compressed by this tendon. The procedure is known as *post-tensioning*, and it is usual subsequently to fill the duct surrounding the cable with grout. A grout of cement, with no more than sufficient water for the workability required, is suitable. Sand is not recommended[1, 2]. Special plasticisers[3, 4] are recommended to give better quality grouting. Air entrainers[4] can be used instead of, or in addition to plasticisers. It is vitally important not to trap pockets of water in ducts, as they have frozen in winter and caused trouble. Soroka and Geddes[5] report 'the ultimate moment and pattern of cracking are hardly influenced by grout quality'. Szilard[6] reports particular concern with regard to the adequacy of the strength of the grout and its corrosion resisting properties. Refs. 5 and 6 list 75 and 103

references respectively on this subject. The author has experience of a special polyester material which would seem to be excellent for strength and workability for use in even damp ducts, though its rapid setting time would be its greatest disadvantage in use and some development in this respect would be necessary. Epoxy resins[3] are affected by water, as in the experience of the author the hardeners react with water.

3. A variation on method 2 is to place the tendon in the sheath before concreting. It is usually easier to thread the tendon in the sheath before concreting than in the duct after concreting. This does not, of course, allow inflatable tubes to be used for forming the duct. The latter method appears to be cheaper from the point of view of forming the duct, but on the whole, in the U.K., when the extra cost of positioning the inflatable tubes and threading the ducts they form is considered, it is usually more economical to place the tendon in the expendable tubing before casting.

4. Another variation on method 2 is to make the concrete member in precast portions which are placed together on the site, the joints between such members being dry packed with cement:sand mortar, usually after the tendons have been threaded through the blocks.

5. Prestressing forces can be exerted on structures in suitable places by jacks. For example, hydraulic jacks have been used in the abutments of dams arched in plan, to exert known forces in favourable directions and achieve economies in the amounts of concrete required in the dams.

8.1.1 Advantages and disadvantages of prestressing

The chief advantages of prestressed concrete are in reducing the quantities of steel and concrete required and in eliminating cracks. The disadvantages are the extra labour costs connected with the stressing of the tendons, and with other items.

Prestressing strengthens a beam in shear and can give a useful saving in shear reinforcement, useful with regard to cost and sometimes especially with regard to facility of detailing. The author has on occasions post-tensioned jointed precast structures solely because of the weakness of the joints in shear.

In the U.K., if a member can be equally well constructed in prestressed or ordinary reinforced concrete, then the latter is usually more economical. When, however, large spans are required with shallow depths, e.g. for bridges, precast floors and so on, and the ordinary reinforced concrete is structurally unacceptable, then prestressed concrete is the only answer in concrete, and, if there is a reasonable repetition in the making of members (to reduce shuttering costs), in the U.K. it is sometimes more economical than structural steelwork. If a factory is highly organised in the manufacture of prestressed flooring units it is sometimes found that units which could be of ordinary reinforced concrete can be made shallower in prestressed concrete and can thus be less expensive overall by making savings in transportation, handling and stacking. When the spans for bridges are sufficiently short to make prestressing cheaper than steelwork, prestressed concrete has the great advantages over steelwork of relative freedom from maintenance, and fire resistance.

Prestressed concrete construction is often more expensive to design than ordinary reinforced concrete work. In post-tensioned *in situ* structures, prestressing procedures have to be carefully planned so that tensioning one cable does not make previously tensioned cables deficient in stress and also does not cause undesirable stresses to develop due to the eccentricity of the prestressing force; this eccentricity will usually be eliminated when the prestressing is satisfactorily completed. Sometimes this involves larger amounts of structures to be shuttered or alternatively supported before prestressing than would be necessary if the structure were of ordinary reinforced concrete. In such circumstances prestressing sometimes slows down the speed of construction and increases the shuttering required for a contract.

Members designed with prestressed concrete can be very flexible and the designer must be careful that deflections, cambers and flexibilities are satisfactory, as some of these factors can be the design criteria.

It is conceivable even to pay more for prestressed concrete structures than for ordinary reinforced concrete structures when resistance to corrosion is important; the life of the prestressed structure will be greater because of the absence of cracks. Such structures as docks, wharfs and jetties which are exposed to sea water, exposed structures at gas works, bridges exposed to pollution, structural work in dairies exposed to lactic acid, are common examples of concrete structures exposed to corrosive elements.

8.2 Materials

Prestressed concrete uses highly stressed steel and concrete, and good materials and workmanship are most important. Failures have occurred due to corrosion of tendons. The concrete and grouting materials must be non-corrosive to the steel and dense for strength and for resistance against water or corrosive liquids endeavouring to come into contact with the steel. Calcium chloride is detrimental in concrete that allows water to contact the steel. Generally the quantities of chlorides and sulphates should be strictly limited in the concrete materials. The corrosion of tendons is due to pitting and hydrogen embrittlement as well as stress corrosion. Ref. 6 is useful.

If high alumina cement is to be used, refer to Section 2.1.

8.2.1 Stress corrosion

This is a very important problem. There have been failures, and there has been much research, particularly in connection with prestressed concrete bridges, and also because the strands used for tendons are also used for cables of suspension and cable-stayed bridges and funicular mountain railways. The author has seen considerable research in progress in stress corrosion in Paris and Zurich. Leonhardt[3] establishes conditions which must exist for stress corrosion of prestressing wires.

8.3 Losses of prestress

The stress initially effected in the tendons is reduced by the following losses:

1. *Relaxation (creep) of steel.* The high stresses used in the tendons mean that the steel is often stressed slightly beyond its limit of proportionality. Hence, after anchorage to the concrete, the strain in the steel can increase slightly due to creep, thus reducing the stress in the prestressed concrete. With pretensioned members, this loss can be greatly reduced by tensioning say in the afternoon and then suitably increasing the strain of the tendons next morning before casting. This is an operation which interferes with progress and increases labour costs, and for overall economy it is usually better not to try to eliminate creep but to consider it as a loss in prestress. A rise of temperature helps the steel to creep and can thus increase creep loss. The relaxation loss depends on the type of tendon and the magnitude of the stress it experiences.

2. *Elastic deformation (strain) of concrete.* When pretensioned wires are released they compress the concrete, the concrete strains, and thus reduces the strain and hence the stress in the wires. This is known as loss of prestress due to strain (or elastic deformation). A

post-tensioned member with only one tendon does not, in theory, experience a strain loss because as the jack strains the tendon it compresses the concrete. When more than one tendon is used, then as each tendon is strained the jack increases the strain in the concrete; this reduces the strain in the tendons already anchored; i.e. strain losses occur in all but the last tendon to be stressed. All these losses total less than those experienced with pretensioned concrete. When pretensioned wires are released they will shorten owing to the concrete becoming strained and stressed (i.e. prestressed). The shortening of the wires divided by their length is the loss in strain of the wires, say ε_1. This shortening must be the same for the concrete immediately in contact with the wires and this is unstressed before the shortening and then stressed due to the shortening. Hence the strain in the concrete is also ε_1. Applying Hooke's law, the loss of stress in the wires is $E_s\varepsilon_1$ and the gain of stress in the concrete is $E_c\varepsilon_1$. Hence loss of stress in wires $= E_s\varepsilon_1 = E_s$ (Stress in concrete/E_c) $= \alpha_e$ (Stress in concrete). For post-tensioning, with several tendons, the resulting loss of stress can be taken as half this amount.

3. *Shrinkage of concrete.* Shrinkage is discussed in Chapter 2. As concrete shrinks after the tendons have been anchored to the concrete, the concrete member shortens and hence so does the tendon, thus releasing some stress in the tendon. In the case of pretensioned concrete, the shrinkage effect begins as soon as the concrete is cast, but with post-tensioned concrete the concrete is able to shrink before the tendon is stressed. If there were no longitudinal reinforcement the shrinkage would be restricted only by friction with moulds, etc., and most of the shrinkage would occur before stressing. Longitudinal reinforcement interferes with this process to some extent. Humidity and temperature also affect shrinkage. For practical design CP 110 gives suitable recommendations for calculating the loss of prestress due to shrinkage of the concrete.

4. *Creep in concrete.* Creep has already been explained in Chapter 2. As the concrete creeps it reduces the strain and hence the stress in the prestressing tendons. With prestressed concrete, creep is not under a constant stress, as considered in Chapter 2, because the stress in the concrete is reducing as the concrete creeps. The creep loss may be estimated by reference to CP 110. The loss is greater for pretensioned than for post-tensioned members. Pretensioned tendons rely upon their bond to the concrete for anchorage and in time this releases (or creeps) slightly; this *creep of bond stress* is not counted as a separate loss, so is accounted for as an increase in creep loss.

5. *Slip of anchorage.* This refers to the tendons losing stress after anchorage due to the anchorage device slipping; e.g. wedges are pulled forward in their jaws as the stress is taken up by the anchorage.

This should be assessed for the particular system used. For prestressing over short distances it is preferable that this allowance should be as small as possible, as the greater the allowance the greater the probable error in the reliability of this quantity. For this reason Wilby[7] has found Lee McCall bars useful for prestressing over short lengths; the relative movement between the nuts and threads of this system causes only very little loss of stress.

6. *Friction in jack and anchorage system.* In pretensioning, if the extension of the wire is measured directly then the friction in the jack is not a loss to be deducted from this prestress measurement. If, however, in pretensioning, and mostly in post-tensioning, the prestress is measured say on the body of the jack by the movement of the movable part relative to the stationary part, then there will be friction in the jack and this frictional force will have been included in our measurement of the force in the wire, by measurement of its extension. Hence a suitable allowance should be made for this frictional loss and this depends on the type of jack used. In post-tensioning systems where the tendons are deviated just before the final anchorage, indeed deviated to effect the final anchorage because of the space required by jacks and anchorages, there is a frictional loss related to the pressure of the tendons on the sides of the deviating device used by the particular system. This must be allowed for in determining the prestressing force in the tendons.

7. *Friction along duct in post-tensioning.* With regard to a straight tendon it will normally be detailed not to touch its duct, but in practice, because of lack of straightness of tendon and duct, there will be some contact between a tendon and its duct. The duct will tend to deviate the tendon, perhaps in some kind of wobble along the duct. Because of this deviation there must be pressure between tendon and duct and thus frictional forces between the two when the tendon is being stressed. When a tendon is taken round a bend, the tendon exerts pressure on the duct, or concrete tank wall, etc., and there is friction associated with this pressure on tensioning the tendon.

CP 110 recommends the following formula which has been justified by tests:

$$P_x = P_0 \exp\left[- Kx - (\mu x/r_{ps})\right] \tag{8.1}$$

which for small values of Kx and $\mu x/r_{ps}$ can be approximated to

$$P_x = P_0\left[1 - Kx - (\mu x/r_{ps})\right] \tag{8.2}$$

where

P_0 = Prestressing force in a tendon at the jacking end.
P_x = Prestressing force at any distance x from the jack.
K is a constant depending on how much the duct is likely to

deviate, i.e. how rigid is the sheath, how often it is supported, how much vibration is used for the concrete.

μ is the coefficient of friction between the tendon and the duct surface, or surface of concrete tank wall, etc.

r_{ps} = Radius of curvature.

e = 2.718.

In the case of tendons which are not to be finally bonded to the member, they can be lubricated, and some tendons can be purchased enclosed in polythene sheaths packed with suitable grease. The author has used the latter on the outside of a dome which required strengthening as an emergency measure—the polythene and grease have good weather resistance.

8. *Steam curing.* This can interfere with the losses due to creep and shrinkage of the concrete, and relaxation of the steel. On a long line pretensioning system where a long rigid mould is employed, with partitions, the steam expands the wires, causing them to lose some stress. Then the concrete hardens. Then if the members cool before release of wires, the stress increases in the wires due to contraction, and as there is friction between the long mould and the line of members in it, a serious tension can develop in the concrete at the two ends of the line of members. This can be avoided by releasing the wires before cooling of the members, or by making sure that the members are free to move, or by using individual movable moulds along the long line system.

8.4 Limit state design of members

Members must be designed for the following, according to CP 110:

1. Limit state of cracking, due to flexure.
2. Ultimate limit state, due to flexure.
3. Prestressing requirements; losses; maximum initial prestress; end-block design or transmission length requirements.
4. Ultimate limit state; shear.
5. Limit state of deflection.
6. Considerations affecting design details.

For an exposed structure it might be required to eliminate cracks completely, whatever the particular loading being experienced by the member. In the U.K. absence of cracks was originally considered to be one of the prime advantages of prestressed concrete. Generally speaking the greater the amount of flexural tension which can be allowed the more economic will be the construction, e.g. less tendons required, but the greater the danger of cracking.

The ultimate strength of a prestressed concrete beam is little different whether the prestressing is applied or not. The greater the amount of prestressing applied to such a beam the more it will be possible to reduce the size of the cracks (they can even be eliminated), the amount of the deflection, and its rigidity (proportional to second moment of area).

For design purposes (with regard to the limit state for cracking) CP 110 suggests three classes of structures thus:

Class 1. No tension is allowed to be taken by the concrete except for a limited amount due to prestress alone.

Class 2. Tension is allowed to be taken by the concrete but the amount is limited to preclude noticeable cracking.

Class 3 refers to *partial prestressing*, where large theoretical tensile stresses are allowed which cannot exist because of exceeding the modulus of rupture of the concrete, but these theoretical tensile stresses are limited so that the cracks which will occur are not likely to allow rainwater to penetrate to the reinforcement, etc.

The designer has to choose his limit state of cracking according to the conditions of exposure, and the quality required, of the structure. If the designer is, say, concerned about temperature stresses due to the member not being perfectly free to move, then he may require no tensile stresses, and he may require the minimum compressive stress at any stage of loading experienced under working conditions to be slightly in excess of the maximum tensile temperature stresses. This would be more conservative than Class 1.

Class 1 would be used say for exposed structures (exposed to polluted atmosphere, sea water, etc.). Class 2 would be used for more economy than Class 1, when durability is not so important. Class 3 would be used for greater economy, but of course one of the advantages of prestressed concrete, absence of cracking, is sacrificed. Class 3 could be suitable where there would be no tensile stresses under most working loads, but yet for the infrequent maximum working load of short duration tensile stresses would be induced in the member. It has been used for some railway bridges.

The sequence of design for Classes 1 and 2 is suggested to be in the order 1–6 (page 144). For Class 3 the sequence is suggested to be 2, 1, 3, 4, 5, 6.

To design for the various limit states of CP 110 it is necessary to be able to calculate stresses and deflections at working loads. This is done with the elastic theory. It is also necessary to design for ultimate strength, using plastic theory. In addition, transmission lengths for prestressing wires and end-block designs for post-tensioned members must all be adequate. All these methods will now be discussed.

In design it is always necessary to find a suitable section and its reinforcement, before all the checks of the adequacy of 1–6 (page 144) are ascertained with adequate precision. With experience the original estimate of the section may need little or no alteration as a result of these checks. For optimisation one would program the procedure so that the computer can keep modifying the original estimate of the section to satisfy the various checks as economically as possible. This is simply a matter of programming the procedures of design which follow in this chapter.

8.4.1 Simple assessment of size of prestressed members

As previously mentioned, experience helps this procedure. One can be guided by observing sizes of members of similar jobs from publications, etc. Alternatively, or in addition, one can choose the type of concrete to be used—one which is not too difficult to achieve with the methods to be used and standard of product required—and proceed as follows.

Example 8.1. An initial estimate is required of a suitable I-shaped cross section for a prestressed concrete beam which has to resist a total bending moment at mid span at working loads of 870 kN m (inclusive of its self-weight), and is to be designed for a limit state of cracking of Class 1.

It is fairly easy for the manufacturer to obtain a concrete of characteristic cube strength at 28 days (when we assume the structure may need to withstand its working load) of 40 N/mm², and this concrete can be made of early enough strength for the requirements at transfer. Referring to Table 32 of CP 110, and because we are considering concrete stresses $\gamma_m = 1.3$, but as an allowance has been made for this in the table, the allowable compressive stress is $0.33 \times 40 = 13.2$ N/mm² $= 13\,200$ kN/m². If the tendons are to be straight then the bending moment due to the weight of the member will reduce the prestressing at mid span, but not at the supports, and thus the supports are the critical sections for deciding the amount of prestressing. At these sections at working loads the prestressing could be as *Figure 8.1(a)*. This would also be the prestressing at the mid-span section, because of the straight tendons. The total bending moment at mid span can therefore give a stress distribution as *Figure 8.1(b)*, which is superimposed upon (a) to give (c) in *Figure 8.1*, assuming the section to have its neutral axis at mid depth of the section. The section can hence be designed as for *Figure 8.1(b)*; thus the section modulus needs to be $870/13\,200 = 0.065\,91$ m².

Try the section shown in *Figure 8.2*. Its second moment of area is $[(0.45 \times 1.1^3)/12] - [(0.3 \times 0.8^3)/12] = 0.037\,11$ m⁴ and its section modulus is therefore $0.037\,11/0.55 = 0.067\,48$ m³.

This section will therefore be suitable. The bending moment used included an allowance for the self-weight of the member. This had to be estimated and should be checked against the section now obtained. If the estimate is found

L

to be wrong then the section we have just designed gives a good clue to a revised estimate of the self-weight of the beam for use in a revised design.

Example 8.2. It might be useful to continue the design of Example 8.1 to assess approximately the tendons required. Suppose the beam is pretensioned and 7 mm diameter wires are to be used. From Table 29 of CP 110 the specified characteristic strength of these wires is 60.4 kN and the cross-sectional area of each wire is 38.5 mm^2. For assessing stresses $\gamma_m = 1$, and the maximum initial prestressing force in a wire, according to CP 110, would normally

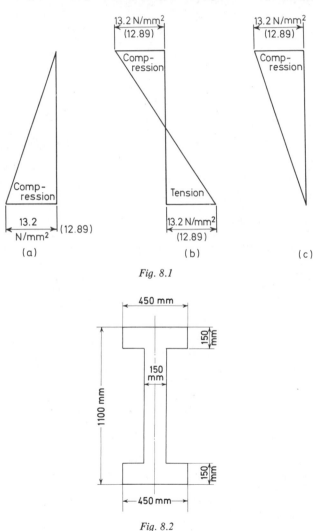

Fig. 8.1

Fig. 8.2

be 70% of this 60.4 kN = 42.28 kN. The ACI-ASCE 323 Report suggests that for approximate purposes total losses can be taken as 245 N/mm² for pretensioning and 175 N/mm² for post-tensioning, but loss due to friction between tendon and duct must be added to the 175. Hence total loss of prestressing force per wire = 0.245 × 38.5 = 9.433 kN and prestressing force per wire after losses = 42.28 − 9.43 = 32.85 kN. The prestressing force required after losses (see *Figure 8.1(a)*), will be the average prestress multiplied by the area of the cross section.

Area of cross section = 0.45 × 1.1 − 0.3 × 0.8 = 0.255 m².

Now section was slightly larger than required. We might as well allow for this, so the stress in *Figure 8.1(a)* now becomes 870/0.067 48 = 12 890 kN/m² = 12.89 N/mm². This figure will therefore be used instead of 13.2 in *Figure 8.1*. It is shown in brackets in the figure.

Prestressing force required after losses when member finally in use, from *Figure 8.1(a)* = 0.5 × 12 890 × 0.255 = 1643 kN. Therefore

$$\text{Number of wires required} = 1643/32.85 = 50.02 = 51 \text{ wires}$$

The designer has to check whether or not these can be placed in the section, with the distances between wires and covers specified by CP 110; and the centroid of the wires should coincide with the centroid of the force calculated from the stress distribution of *Figure 8.1(a)* and the cross-sectional areas of *Figure 8.2*. If this is not possible, then larger tendons may be satisfactory, but if unsatisfactory the designer starts again with another size of section.

In this case the wires can be accommodated in the section. They will mostly be placed in the bottom flange, perhaps two or more in the top flange and perhaps a few in the web.

From *Figure 8.1(a)* the bending moment due to the prestressing force

$$= 1643 \times e = (12\,890/2) \times 0.067\,48$$

where e is the depth of the resultant prestressing force below the centre of the depth of this symmetrical section. Therefore $e = 0.2647$ m.

As mentioned before, the wires have to be disposed so that their centroid is at this depth.

8.4.2 Assumptions for elastic design

Of the following assumptions, 1–4 are the same as those described in Chapter 3:

1. Plain sections subjected to bending remain plane after bending.
2. Stress is proportional to strain for both the steel and the concrete.
3. Perfect bond is assumed between the steel and the concrete. In the case of post-tensioning this theoretically applies after the tendon has been grouted.

4. Depths of reinforcements relative to the depth of the concrete member are considered to be negligible.

5. Allowances must be made for shrinkage and creep losses.

6. Young's modulus for concrete is the same in tension as compression; this is reasonably true.

8.4.3 Limit states of stresses and deflections

During the life of a prestressed concrete beam there are many changes in the stresses and deflections it experiences, and all the worst possibilities should be investigated. When all this has been evaluated, if anything is wrong then one has to return to the beginning and re-estimate the size of the section. Hence one has to concentrate on the most likely worst cases first, so that if re-design is necessary one finds this out as soon as possible.

Essentially a member has to be designed for stresses at *transfer* of prestress from tendons to concrete. This is an important limit state, as the concrete is often not very old and hence not as strong as it will be when in the final structure; also, the tendons have not experienced losses as great as they will experience in the final structure.

Then the member, if not *in situ*, will be handled, stacked, loaded, transported, unloaded, perhaps stacked and then lifted into position. All these operations, if not skilfully performed, could impose many adverse stresses. It is usually best, for prestressed concrete, to have spreaders for slings of cranes and to use lorries with long backs so that beams are always supported at their ends as they have been at transfer, and will be in the final structure. Then adverse stresses can be eliminated, and there is no need to design for this limit state of handling, transportation, erection, etc.

A member must also, of course, be designed for its limit states of stresses and deflection when in its final position in the structure.

8.4.4 Simplified elastic design of prestressed concrete beams

The simplification is by way of ignoring the steel reinforcements in calculating the cross-sectional area, depth of neutral axis and second moment of area of the concrete section. This reduces the work of the calculations considerably, as for very accurate calculations various different sectional properties are required. For example, when pretensioning, the areas of the wires and the concrete they displace should be included in the calculations of the sectional properties and different modular ratios should be used for transfer and final serviceability. For post-tensioning, at transfer the tendon and duct should be excluded, but any other steel included in calculating sectional properties. On the other hand, for limit state of serviceability (i.e. after the duct has been grouted) all the concrete (including grout), tendons and any other reinforcements should be included in the calculations of sectional properties, these reinforcements having a different modular ratio to that used at transfer.

This simplified method is adequate for many purposes, as the percentage of steel in the cross section is generally low enough to cause little error, and this error tends to cause excess safety.

The losses are firstly taken as percentages of the initial prestressing tendon forces. This enables the concrete stresses to be obtained and then the losses can be obtained more accurately from these stresses. If it is then found that the original estimate of losses was not good enough, an adjustment is made and the design repeated. The process can be repeated until the designer is satisfied—it can quite simply be programmed for a computer. However, with experience a designer often does not need to alter his first estimate, as he will have slightly overestimated so as not to have the trouble of re-design; the computer is of course useful for optimisation here. The method is illustrated in the following examples.

Example 8.3. Continue the design of the beam of Example 8.1. Having approximately checked the stresses it might now be best approximately to check the limit states of deflection in case we have to alter the section on this count. When we are interested in the maximum deflection in service, the concrete then has a characteristic strength of 40 N/mm^2, and γ_m for concrete is unity, so from Table 1 of CP 110 $E_c = 31$ kN/mm^2. Shrinkage and creep have been allowed for in the losses assumed. When we consider deflection at transfer we will assume that the concrete has a characteristic strength of 30 N/mm^2, and γ_m for concrete is unity, so from Table 1 of CP 110, $E_c = 28$ kN/mm^2.

Assuming the beam is simply supported over a span of 22 m and that all loading is uniformly distributed (q), then $(q/8) \times 22^2 = 870$, $\therefore q = 14.38$ kN/m.

From CP 110, Clause 2.2.3, the limit states of deflection are as follows:

1. If finishes are to be applied the span-to-total-upward-deflection ratio should exceed 300. This refers chiefly to floor and roof units which can have varying cambers, often because of releasing the wires when the concrete is not strong enough on the prestressing beds—the indicative cubes are sometimes compacted very much more thoroughly and sometimes cured more favourably than most of the concrete in a member and, under these bad circumstances, are misleading. The less the upward deflection, the less the problem and hence this limitation suggestion of CP 110.

At transfer the losses will not be as great as finally. For pretensioning they can be very approximately 10–15% (assuming the relaxation losses of the steel are kept modest). Supposing we take 10% to be on the safe side. At transfer the smaller losses give greater concrete stresses, which are usually the most limiting consideration at transfer. (Note that a safe and not excessively conservative figure for post-tensioned concrete would be just the steel relaxation loss if the tendons are stressed simultaneously and there are no excessive losses due to severe curvature, such as for a circular tank or dome.) Prestressing force of 1643 kN was based on losses of $(9.43/42.28) \times 100 = 22.3\%$. Then at transfer prestressing force after losses = $1643 \times (100 - 10)/(100 - 22.3) = 1903$ kN. The bending moment due to this prestressing force = $1903 \times 0.2647 = 503.7$ kN m. The deflection upwards due to this constant bending moment (the span on the prestressing bed is the overall length of the beam, say 23 m) = $(503.7 \times 23^2)/(8 \times 28 \times 10^6 \times 0.037\,11) = 0.0321$ m. This is reduced by the downwards deflection due to the self-weight of the beam, which is $(5 \times 6.018 \times 23^4)/(384 \times 28 \times 10^6 \times 0.037\,11) = 0.0211$ m, where the self-weight of the beam, assuming the weight density of prestressed concrete is 23.6 kN/m^3 (mass density of prestressed concrete = 2400 kg/m^3), is $0.255 \times 23.6 = 6.018$ kN/m. At transfer, therefore, the total upward deflection = $32.1 - 21.1 = 11.0$ mm.

This gives a span-to-deflection ratio of 2091, which is greater than 300 and therefore satisfactory should the member be used in this way. (In this particular example it was not necessary to calculate the 21.1, as the 32.1 without the reduction of 21.1 would still have been satisfactory for the span-to-deflection ratio of 300, but this will not always be the case.)

2. The final span-to-deflection ratio should exceed 250, the deflection being measured below the level of the supports. In the present example the deflection downwards = $(5 \times 14.38 \times 22^4)/(384 \times 31 \times 10^6 \times 0.03711) = 0.038\,13$ m. The bending moment due to the prestressing force = $1643 \times 0.2647 = 434.9$ kN m. The deflection upwards due to this constant bending moment = $(434.9 \times 22^2)/(8 \times 31 \times 10^6 \times 0.037\,11) = 0.022\,87$ m. Hence the deflection below the supports is $38.13 - 22.87 = 15.26$ mm. This gives a span-to-deflection ratio of 1442, which is satisfactory as it exceeds 250.

3. Partitions and finishes, either above or below, if the beam is in a building, can be damaged by excessive deflections. CP 110 generally suggests limiting the deflection to 20 mm and to a span-to-deflection ratio greater than 350. These deflection calculations are for deflections after the fixing of the partitions and the applications of the finishes. We can therefore assume the concrete to be at least 28 days old and we are essentially interested in the subsequent deflection due to live load. However, if say glass partitions are built up to the soffit of the beam, then no live load deflection is tolerable and details have to be devised to, for example, allow a beam to slide past rather than bear on to a partition.

In the present example, the self-weight of the beam, from 1, is 6.018 kN/m, hence the live load is $14.38 - 6.018 = 8.362$ kN/m and the deflection due to this = $(8.362/14.38) \times 38.13 = 22.17$ mm. The span-to-deflection ratio is 992.3, which is greater than 350 and therefore satisfactory.

It is, however, greater than 20 mm and therefore not as recommended by CP 110. If this is acceptable practically then the design does not need revision to reduce this deflection. Thus in the present example the beam might not be suitable in a building.

Example 8.4. Before we examine in more detail the preliminary design given in Examples 8.1 and 8.2, it would be advisable to determine approximately the adequacy of stresses at transfer of this design.

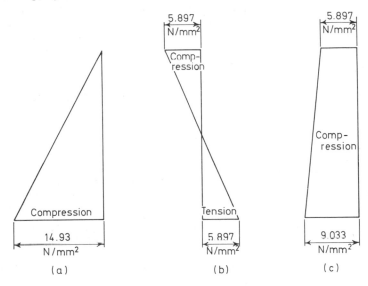

Fig. 8.3

At transfer the losses will not be as great as finally. They can be 10–15%. Supposing we take 10% to be on the safe side. At transfer less losses give greater concrete stresses, which are usually the most limiting consideration at transfer. Prestressing force of 1643 kN was based on losses of 22.3%. Hence, referring to *Figure 8.1(a)*, the stress of 12.89 N/mm², which is directly proportional to the prestressing force, will be altered for conditions at transfer pro rata to the different prestressing forces at transfer and finally, and it thus becomes $12.89 \times (100 - 10)/(100 - 22.3) = 14.93$ N/mm². *Figure 8.3(a)* therefore shows the distribution of prestress at transfer at the supports. At transfer the member will usually hog upwards and hence the mid-span section withstands the maximum bending moment due to the self-weight of the member superimposed upon the prestress at this section. For calculating this bending moment we should use the overall length (23 m) of the beam. The maximum bending moment due to self-weight $= (6.018 \times 23^2)/8 = 397.9$ kN m, and the extreme fibre stresses due to this bending moment $= 397.9/0.067\,48$ kN/m² $= 5.897$ N/mm². *Figure 8.3(b)* therefore shows the distribution of stress at mid span due to the self-weight loading. Algebraically adding these stresses to the prestress shown in *Figure 8.3(a)* we obtain *Figure 8.3(c)*, which gives the resultant distribution of stress at mid span at transfer. Referring to Table 36 of CP110, the concrete strength at transfer will need to be the greater of $14.93/0.5 = 29.86\,\text{N/mm}^2$ or $9.033/0.4 = 22.58\,\text{N/mm}^2$. This agrees with our assumption of $30\,\text{N/mm}^2$ in Example 8.3.

Example 8.5. In Examples 8.1, 8.2, 8.3 and 8.4 we have made an approximate design of a prestressed concrete beam. We shall now check for this beam the limit states determined by elastic theory and concerning stresses and losses.

For these limit states CP110 gives $\gamma_m = 1$ for steel and 1.3 for concrete. Considering the losses:

1. *Relaxation of steel.* CP110 refers us to BS 2691, 1969, and supposing we use cold drawn and prestraightened low relaxation wire, then, as the initial prestress we took is 70% of characteristic strength, Table 6 of this British Standard gives the maximum percentage relaxation after 1000 h as 2%.

2. *Elastic deformation of concrete*

   (a) At transfer

   Support: stress in concrete at level of centroid of wires (from *Figure 8.3(a)*) = $\{(550 + 264.7)/1100\} \times 14.93 = 11.06\,\text{N/mm}^2$.

   From Example 8.3, $E_c = 28\,\text{kN/mm}^2$. Clause 2.4.2.4 of CP110, for the wires, gives $E_s = 200\,\text{kN/mm}^2$. Hence $\alpha_e = 200/28 = 7.14$. Therefore, as explained earlier, loss of prestress = $7.14 \times 11.06 = 79.0\,\text{N/mm}^2$. Using cross-sectional area given in Table 29 of CP110, the loss of force per wire = $79.0 \times 38.5\,\text{N} = 3.042\,\text{kN}$. Hence the percentage loss of initial prestressing force = $(3.042/42.28) \times 100 = 7.19\%$.

   Mid span: at level of centroid of wires, stress due to self-weight = $5.897 \times (264.7/550) = 2.838\,\text{N/mm}^2$. Therefore resultant stress at this level due to self-weight and prestress = $11.06 - 2.838 = 8.22\,\text{N/mm}^2$. Therefore, as previously, percentage loss of initial prestressing force = $7.19 \times (8.22/11.06) = 5.34\%$.

   (b) In service

   Support: stress in concrete at level of centroid of wires (from *Figure 8.1(a)*) = $\{(550 + 264.7)/1100\} \times 12.89 = 9.55\,\text{N/mm}^2$. From Example 8.3, $E_c = 31\,\text{kN/mm}^2$. Clause 2.4.2.4 of CP110 for the wires gives $E_s = 200\,\text{kN/mm}^2$. Hence $\alpha_e = 200/31 = 6.452$. Therefore loss of prestress = $6.452 \times 9.55 = 61.61\,\text{N/mm}^2$. Therefore percentage loss of initial prestressing force = $7.19 \times (61.61/79) = 5.61\%$.

   Mid span: from *Figure 8.1(c)* stress at level of wires = $12.89 \times (550 - 264.6)/1100 = 3.344\,\text{N/mm}^2$. Therefore percentage loss of prestress = $5.61 \times (3.344/9.55) = 1.96\%$.

3. *Shrinkage of concrete.* Supposing the beam is cured in effect in water— say by covering with wet hessian cloth which is covered with polythene sheet; there will then be no shrinkage loss at transfer. Table 41 of CP110 gives concrete shrinkage values for pretensioning between three and five days after casting, curing at exposures of 90% and 70% r.h. respectively as far as transfer. Let us assume that we cure after transfer at normal exposure until it is in use, then, guided by *Figure 2.5*, take the maximum shrinkage per unit length as 0.04%. Taking the shortening movement of the tendon as the same

as the concrete shrinkage, then strain loss in tendon due to shrinkage $= 400 \times 10^{-6}$. Hence corresponding loss of stress $= 400 \times 10^{-6} \times 200 \times 10^3 = 80 \text{ N/mm}^2$. Therefore percentage loss of prestress (finally) $= (80/79) \times 7.19 = 7.28\%$.

4. *Creep of concrete.* At transfer, creep has had negligible time to take place, hence we take this loss as zero. The stress in the concrete at transfer will cause subsequent creep, and it is upon this stress that the final creep is based. Referring to Clause 4.8.2.5 of CP 110, as cube strength at transfer is 30 N/mm², creep per unit length is $48 \times 10^{-6} \times (40/30) = 64 \times 10^{-6}$ per N/mm². According to CP 110, if the maximum stress at transfer exceeds $\frac{1}{3} \times$ Cube strength at transfer $= \frac{1}{3} \times 30 = 10 \text{ N/mm}^2$ then the creep loss should be increased. At transfer *Figure 8.3(a)* gives the stresses at each support and *Figure 8.3(c)* gives the stresses at mid span. At mid span the stresses do not exceed 10 N/mm² so the creep loss is satisfactory. At each support, as the maximum stress is approximately half the cube strength, the creep per unit length from CP 110 is $1.25 \times 64 \times 10^{-6} = 80 \times 10^{-6}$ per N/mm². The stress causing creep will depend upon whether the beam is supporting its own weight only or its full load most of its life. At the level of the centroid of the wires the former gives a stress of 11.06 N/mm² at support and 8.22 N/mm² at mid span, whilst the latter, from *Figure 8.3(a)* and *Figure 8.1(b)*, gives 11.06 N/mm² at support and $11.06 - (264.7/1100) \times 12.89 = 7.96 \text{ N/mm}^2$ at mid span. Supposing the imposed load is rarely applied, so that we take the worst of the cases just mentioned. When in use therefore the creep per unit length is (a) $11.06 \times 80 \times 10^{-6} = 885 \times 10^{-6}$ at the support, and (b) $8.22 \times 64 \times 10^{-6} = 526 \times 10^{-6}$ at mid span. As the movement of the concrete is assumed to be the same as that of the tendon, then the loss of stress in the tendon is (a) $885 \times 10^{-6} \times 200 \times 10^3 = 177 \text{ N/mm}^2$ at the support and (b) $526 \times 10^{-6} \times 200 \times 10^3 = 105.2 \text{ N/mm}^2$ at mid span. These can be expressed as (a) $(177/61.61) \times 5.61 = 16.12\%$ at support, and (b) $(105.2/61.61) \times 5.61 = 9.58\%$ at mid span.

5. *Slip of anchorage.* Suppose the wedges at each end pull in 3 mm and our system is one where we jack the movable wire anchorage block away from the prestressing bed, which has a length of say 75 m; then this loss, if not allowed for when stressing, would be $(6/75\,000) \times 200 \times 10^3 = 16 \text{ N/mm}^2$. But we will allow for this when stressing and extend the movable anchorage block 6 mm more than its required amount.

6. *Friction in jack and anchorage system.* This is nil because of the way we are pretensioning; see 5 and also Section 8.3, para. 6.

Summarising the losses: at transfer total loss at mid span $= 2(1) + 5.34(2) = 7.34\%$ and at a support $= 2(1) + 7.19(2) = 9.19\%$; finally, in use total loss at mid span $= 2(1) + 1.96(2) + 7.28(3) + 9.58(4) = 20.82\%$, and at a support $= 2(1) + 5.61(2) + 7.28(3) + 16.12(4) = 31.01\%$.

At transfer we took the losses as 10%, so for greater accuracy we could now try 9.1% for support sections and 7.3% for mid-span sections and repeat the above design. Further such repetitions can then be made until the desired degree of accuracy is achieved. When in use we took the losses as 22.3%. For greater accuracy we would repeat the above design and use losses of 20% for mid span and 31% for support sections. Also for more accurate design, we would repeat the example, considering losses for the wires at their respective levels. We have considered them all as though concentrated at their centroid and this is slightly erroneous. For normal purposes our present accuracy in this problem could be considered satisfactory and hence our design is justified.

8.4.5 *Ultimate limit state due to flexure (bonded tendons)*

If a member has been designed as shown previously, then if the tendons are arranged so that most have a reasonably generous effective depth, checking the ultimate limit state is almost a formality. The exception to this is in the case of CP 110, Class 3, structures (partially prestressed)—see Section 8.4. These could be designed for ultimate limit state first, then deflections at working loads checked before checking the stress systems at working loads and transfer.

In the case of a rectangular beam with one tendon, this is required at about $\frac{1}{3}$ of the height of the beam. Hence when cracking occurs due to overloading the effective depth of this tendon is small, so the tendon does not control the crack widths very well at the soffit. In a case like this the ultimate limit state might not be satisfactory so additional non-prestressed reinforcement might be used and placed as near to the soffit as possible. Likewise in the case of a pole of circular cross section, when the bending moment can be in any direction; if the tendons are arranged around the periphery then the ultimate limit state will most probably be all right, but not if there is say just one tendon down the centre.

As in Chapter 3 the equivalent rectangular stress block due to C. S. Whitney is favoured[8] for predicting ultimate resistance moments, i.e. $f_{cm} = (\alpha/2\beta)f'_c$ and $x_1 = 2\beta x$ where (taking $f'_c = 0.84f_{cu}$) $\alpha = 0.72$ for $f_{cu} \leqslant 33 \text{ N/mm}^2$ and decreases by 0.04 for every 8.21 N/mm² above 33 N/mm², and $\beta = 0.425$ for $f_{cu} \leqslant 33 \text{ N/mm}^2$ and decreases by 0.025 for every 8.21 N/mm² above 33 N/mm². Thus for $f_{cu} \leqslant 33 \text{ N/mm}^2$, $f_{cm} = 0.85 f'_c = 0.714 f_{cu}$ and $x_1 = 0.85x$. It is generally used in the U.S.A. and is the basis of the CP 110 simplified method and many other codes internationally. Towards failure in bending a prestressed concrete beam cracks and behaves like a non-prestressed reinforced concrete beam apart from:

1. The strain in the tendon was not zero at zero loading, as in the reinforcement of the reinforced concrete beam. At zero loading the strain ε_p in a tendon corresponds to the force in the tendon after losses (i.e. the losses which have occurred up to the time of loading to failure) divided by the cross-sectional area of the tendon and its

Young's modulus. Strain in the tendon caused by the loading adds to ε_p. The strain due to the prestress in the concrete can be ignored, as it is negligible compared to ε_p and the strains at failure.

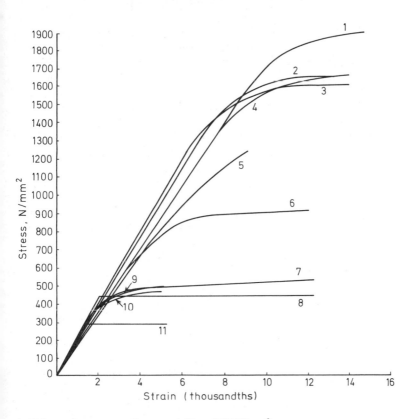

1. 12.7 mm dia. super quality strand, $E_s = 176 \text{ kN/mm}^2$
2. 15.2 mm dia. drawn strand, $E_s = 192 \text{ kN/mm}^2$
3. 5 mm dia. crimped prestressing wire, $E_s = 200 \text{ kN/mm}^2$
4. 15.2 mm dia. French strand, $E_s = 178 \text{ kN/mm}^2$
5. 28.6 mm dia. strand, $E_s = 169 \text{ kN/mm}^2$
6. 32 mm dia. prestressing alloy bar, $E_s = 175 \text{ kN/mm}^2$
7. 16 mm dia. round cold worked high yield reinforcing bar, $E_s = 200 \text{ kN/mm}^2$
8. 20 mm dia. round hot rolled high yield reinforcing bar, $E_s = 213 \text{ kN/mm}^2$
9. 9.5 mm square twisted high yield reinforcing bar, $E_s = 198 \text{ kN/mm}^2$
10. 11 mm square twisted with chamfered edges high yield reinforcing bar, $E_s = 208 \text{ kN/mm}^2$
11. 20 mm hot rolled mild steel reinforcing bar, $E_s = 208 \text{ kN/mm}^2$

Apart from 4, all the above are British products.

Fig. 8.4

2. The stress–strain relationships for tendons are different to those for a reinforcement bar (see *Figure 8.4*). The ultimate resistance moment for an under-reinforced prestressed beam can be obtained as in Example 3.15 for under-reinforced sections, provided the ultimate tensile strength (stress) of the tendons is used for f_s in equations 3.60 and 3.62. To determine if the section is under-reinforced we need to calculate the maximum concrete strain at steel failure (when the steel is at its ultimate breaking strain ε_{su} which is obtained experimentally, i.e. from curves as in *Figure 8.4*) to check that this is less than the maximum known to be possible from experiments, namely 0.003 according to Whitney (0.0035 is used by CP 110). *Figures 8.5(a)* and (*b*) show the distribution of stress across the cross section and the corresponding distribution of strain, respectively.

From similar triangles in *Figure 8.5(b)*, the maximum concrete strain

$$= \varepsilon_c = (\varepsilon_{su} - \varepsilon_p)\, x/(d - x) \tag{8.3}$$

If this is greater than 0.003 (Whitney) then it is an over-reinforced prestressed concrete beam and its ultimate resistance moment

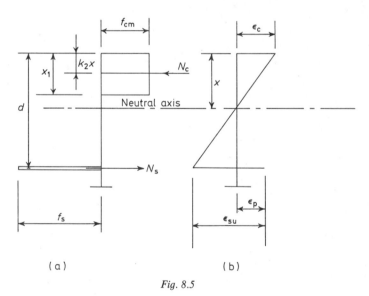

Fig. 8.5

cannot be assessed as previously, as the concrete will disintegrate before the steel reaches its ultimate tensile strength (and corresponding strain ε_{su}). For an over-reinforced section a simple direct calculation cannot be made because the stress in the steel at failure is less than its ultimate tensile strength and is not known initially, hence x_1 cannot be immediately obtained; also, the stress–strain curve for the steel cannot be represented by a simple mathematical expression. A simple solution is by successive approximations (a method suitable for the digital computer). A value of x is assumed and x_1 is obtained from x as before (Whitney). Then equating longitudinal forces

$$f_{cm}A_c = A_s f_s \tag{8.4}$$

where A_s is the cross-sectional area of the tendons, f_s the stress in the tendons at failure, f_{cm} the mean concrete stress of the equivalent stress block, and A_c the area of concrete (cross section can be of any shape) subjected to f_{cm}. This gives f_s and the corresponding strain ε_s is obtained from the stress–strain curve for the tendon. Then from equation 8.3 but substituting $\varepsilon_c = 0.003$ (Whitney) and $\varepsilon_{su} = \varepsilon_s$

$$0.003 = (\varepsilon_s - \varepsilon_p) x/(d - x) \tag{8.5}$$

which now gives x. Now if this disagrees with the value assumed, the calculation is repeated until it is correct. When we are satisfied, taking moments about the line of action of N_c

$$M_u = A_s f_s (d - k_2 x) \tag{8.6}$$

In the case of a rectangular beam, using Whitney's theory, $k_2 x = 0.5x_1$.

Example 8.6. A beam of rectangular cross section 0.25 m wide by 0.45 m deep is post-tensioned by one 25 mm diameter bar at 75 mm above its soffit. The duct enclosing the bar is grouted. Determine the ultimate resistance moment

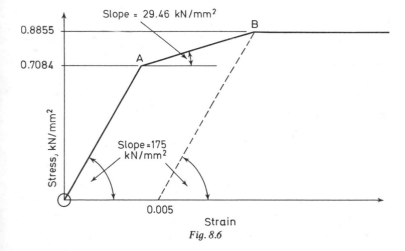

Fig. 8.6

of the section using the simplified CP 110 method and assuming $f_{cu} =$ 40 N/mm^2, γ_m for steel $= 1.15$, stress in 25 mm diameter tendon after losses $=$ 570 N/mm^2.

Equating longitudinal forces (equation 3.60, *Table 3.2*, and CP 110, Table 31) $0.25 \times x \times 0.4 \times 40\,000 = 500/1.15$, $\therefore x = 0.1087$ m. *Figure 8.6* is prepared from CP 110, Fig. 3 and Table 31, $f_{pu}/\gamma_m = 500/491/1.15 = 0.8855$ kN/mm^2 and the minimum strain for the maximum stress to be realised $= \varepsilon_{su} = 0.005 + 0.8855/175 = 0.010\,06$. Also $\varepsilon_p = 0.57/175 = 0.003\,26$. Hence from equation 8.3

$$\varepsilon_c = (0.010\,06 - 0.003\,26) \times 108.7/(450 - 75 - 108.7) = 0.002\,776$$

This is less than 0.0035, so the section is under-reinforced. Hence using equation 8.6

$$M_u = (500/1.15)(0.45 - 0.075 - 0.5 \times 0.1087) = 139.4\,\text{kN m}$$

Example 8.7. Repeat Example 8.6, only with two bars instead of one, both at the same level.

Equating longitudinal forces gives $x = 2 \times 0.1087 = 0.2174$ m. From equation 8.3

$$\varepsilon_c = (0.010\,06 - 0.003\,26) \times 217.4/(450 - 75 - 217.4) = 0.009\,38$$

This is greater than 0.0035, hence section is over-reinforced. Assume $x = 0.1923$ m $= x_1$ (CP 110). From equation 8.4 and CP 110, Table 31, $0.25 \times 0.1923 \times 0.4 \times 40\,000 = 2 \times 491 \times f_s$, $\therefore f_s = 0.7833$ kN/mm^2.

As before $f_{pu}/\gamma_m = 0.8855$ kN/mm^2, so $0.8 f_{pu}/\gamma_m = 0.7084$ kN/mm^2. Hence from *Figure 8.6*, $\varepsilon_s = (0.7084/175) + (0.7833 - 0.7084)/29.46 = 0.006\,59$. Then from equation 8.3 but substituting $\varepsilon_c = 0.0035$ (CP 110) and $\varepsilon_{su} = 0.006\,59$

$$0.0035 = (0.006\,59 - 0.003\,26)x/(0.45 - 0.075 - x) \therefore x = 0.1923\,\text{m}$$

This is in order, the estimate of x being correct. Normally several attempts would be required. Although this trial and error method is favoured by others, and by the writer when using real stress–strain curves for the tendons, direct calculation is to be preferred when using the simplified stress–strain curves of CP 110. To illustrate this: instead of assuming x as before, assume that the strain in the tendons is in the range AB of *Figure 8.6*. Then from equation 8.4 and CP 110, Table 31

$$0.25 \times x \times 0.4 \times 40\,000 = 2 \times 491 f_s \therefore f_s = 4.073x$$

From *Figure 8.6*, $\varepsilon_s = (0.7084/175) + (4.073x - 0.7084)/29.46 = 0.1383x - 0.02$. Using this for ε_{su} and $\varepsilon_c = 0$. in equation 8.3

$$0.0035 = (0.1383x - 0.02 - 0.003\,26)x/(0.375 - x)$$

$$\therefore x = 0.1923\,\text{m and } f_s = 4.073x = 0.7833.$$

Hence f_s does lie in range AB and the calculation is satisfactory. If f_s had been in range AO then one would assume it in this range and make a similar but simpler calculation to the previous one. Then from equation 8.6

$$M_u = (0.7833 \times 2 \times 491)(0.375 - 0.5 \times 0.1923) = 214.5\,\text{kN m}$$

Example 8.8. Repeat Example 8.6 using CP 110, Table 37.

Using Table 31, $f_{pu}A_{ps}/(f_{cu}bd) = 500/(40\,000 \times 0.25 \times 0.375) = 0.1333$. Therefore from Table 37, $f_{pb}/0.87f_{pu} = 1.0$, and $x = 0.290 \times 0.375 = 0.1088$ m.

From CP 110, equation 44 and Table 31, $M_u = 0.87f_{pu}A_{ps}(d - 0.5 \times 0.1088) = 0.87 \times 500 \times 0.3206 = 139.5$ kN m.

Example 8.9. Repeat Example 8.7 using CP 110, Table 37.

Using Table 31, $f_{pu}A_{ps}/(f_{cu}bd) = 2 \times 500/(40\,000 \times 0.25 \times 0.375) = 0.2667$.

For larger amounts of tendons, Table 37 empirically assumes slightly less reliance on the grouting of post-tensioned tendons. Thus to compare with Example 8.7 we should take the figures for pretensioning in Table 37. Thus $f_{pb}/0.87f_{pu} = 1.0$, and $x = 0.580 \times 0.375 = 0.2175$ m. From CP 110, equation 44 and Table 31, $M_u = 0.87f_{pu}A_{ps}(d - 0.5 \times 0.2175) = 0.87 \times 1000 \times 0.27 = 231.6$ kN m.

Using the post-tensioning suggestions of Table 37, $f_{pb}/0.87f_{pu} = 0.883$, and $x = 0.511 \times 0.375 = 0.1916$ m.

From CP 110, equation 44 and Table 31, $M_u = 0.883 \times 870 \times (0.375 - 0.5 \times 0.1916) = 214.5$ kN m.

8.4.6 Additional untensioned steel (bonded tendons)

If the ultimate resistance moment is inadequate, and the other limit states satisfactory, sometimes extra untensioned steel is added. This has negligible effect on the other limit states, and thus saves re-design. This extra steel might be extra prestressing tendons which are not stressed, or reinforcement bar. This steel is placed with maximum effective depth.

Example 8.10. Repeat Example 8.7 but add two 12 mm diameter bars, with 25 mm concrete cover, in the bottom of the beam. Assume $f_y = 460$ N/mm² for these bars.

Equating longitudinal forces, using *Table 3.2*, $0.25 \times x \times 0.4 \times 40\,000 = 2 \times 500/1.15 + 226 \times 0.46/1.15$ ∴ $x = 0.24$ m.

For tendons, from Example 8.6, $\varepsilon_{su} = 0.010\,66$ and $\varepsilon_p = 0.003\,26$. Hence from equation 8.3, $\varepsilon_c = (0.010\,06 - 0.003\,26) \times 0.24/(0.375 - 0.24) = 0.0121$.

Using distribution of strain diagram and similar triangles, strain in 12 mm diameter bars $= 0.0121 \times (419 - 240)/240 = 0.009\,025$, so that maximum stress can be realised in these bars (see CP 110, Fig. 2), i.e. strain greater than $0.002 + 460/(1.15 \times 200\,000) = 0.004$. As ε_c is greater than 0.0035, section is over-reinforced. Equating longitudinal forces

$$0.25 \times x \times 0.4 \times 40\,000 = 2 \times 491 \times f_s + 226 \times 0.46/1.15$$

$$\therefore f_s = 4.073x - 0.092\,06$$

From *Figure 8.6*, $\varepsilon_s = (0.7084/175) + (f_s - 0.7084)/29.46 = 0.1383x -$

0.023 12. Using this for ε_{su} and $\varepsilon_c = 0.0035$ in equation 8.3, $0.0035 = (0.1383x - 0.023\,12 - 0.003\,26)x/(0.375 - x)$

$$\therefore x = 0.2105 \text{ m, and } f_s = 0.7654 \text{ kN/mm}^2$$

As f_s lies in the range AB in *Figure 8.6*, the calculation is satisfactory. We have assumed the strain in the 12 mm diameter bars is large enough for them to develop their maximum stress. Strain in bars $= (0.0035/210.5) \times (450 - 25 - 6 - 210.5) = 0.003\,467$.

Referring to CP110, Fig. 2, this strain is less than 0.004, so the maximum stress is not quite realised. If we reassess this design stress f_{s1}, then repeat the calculation, we should improve the result. If this is done a few times the accuracy becomes adequate. Then M_u is determined by taking moments about the line of action of N_c.

$$M_u = 2 \times 491 \times f_s(0.375 - 0.5x) + 226 \times f_{s1} \times (0.419 - 0.5x)$$

Had the 12 mm diameter high yield bars been replaced by bars of an equivalent strength in mild steel then the strain would have only needed to have exceeded $0.002 + 250/(1.15 \times 200\,000) = 0.003\,087$ (CP110, Fig. 2) for its design yield stress to have been realised, i.e. mild steel for additional unprestressed steel is likelier to simplify the design. In the above design, because the additional steel was not fully stressed a certain amount of trial and error is used, but convergence is very rapid. Some might prefer to guess x to avoid solving the quadratic equation and continue by trial and error as indicated in Example 8.7, but this is slower to converge.

8.4.7 Compression steel

Wires or handling reinforcement placed in the top of a beam are usually inadequately anchored against buckling (see CP110) to be included in the ultimate resistance moment calculations. If compression steel is to be included in these calculations, it is included in the previous calculations in the same way as given in Chapter 3.

8.4.8 Ultimate limit moment due to flexure (unbonded tendons)

In this instance Sections 8.4.5–8.4.7 apply, except that CP110 reduces the force which can be developed in the tendons. The problem is that as loading is applied, instead of the force imposed in the tendon decreasing towards the support as with bonded tendons or reinforcement bars, the force in the tendon is always the same from end to end for an unbonded tendon. Towards failure the first crack occurs at the position of maximum bending moment. At this crack, instead of the tendon being highly stressed locally and anchored on either side of the crack so that its extension is limited (as would be the case if the tendons were bonded to the concrete), when the tendon is unbonded, this high stress extends along its whole length. The whole length thus extends *pro rata* and the extension is considerable, allowing the first crack to open excessively (few if any extra cracks form towards failure), precipitating earlier failure than occurs with a beam with a bonded tendon.

The normal theories treat post-tensioning as if it were pretensioning and just modify the ultimate resistance moment for unbonded tendons as previously and bonded tendons as described in Example 8.9. However, the problem is basically different at pretensioning, working loads and ultimately. Ref. 9 deals at length with this problem at tensioning and at working loads. It takes account of pressures between tendons and their surrounding concrete, which can give high stress concentrations in the concrete[10]. (A failure has been reported where these pressures were considered to be too high and the concrete was under-strength.) Tendons cannot be deflected say vertically by calculate stresses and deflections for beams with tendons of various profiles.

Example 8.11. Repeat Example 8.7, only assuming the tendons are unbonded.

See Example 8.9 but using CP 110, Table 38, instead of Table 37, and supposing $l/d = 20$. From Example 8.6, $f_{pe} = 570 \text{ N/mm}^2$, thus, using Table 31,

$$\frac{f_{pe} A_{ps}}{f_{cu} bd} = \frac{0.57 \times 2 \times 491}{40\,000 \times 0.25 \times 0.375} = 0.1493$$

From Table 38, $f_{pb}/f_{pe} = 1.20$ and $x = 0.46d$. Then using CP 110, equation 44,

$$M_u = 1.2 \times 0.57 \times 2 \times 491\,(0.375 - 0.5 \times 0.46 \times 0.375) = 193.9 \text{ kN m}$$

Compare this result with the 214.5 kN m for bonded tendons in Example 8.9.

8.4.9 Prestressed columns

It is rarely economical or necessary to prestress columns. One example of prestressing columns is in the case of large span pitched-roofed portal frames; in this instance, however, the columns experience very small direct stresses relative to the bending stresses.

8.4.10 Prestressed ties

Prestressed ties are often extremely useful for space frames, arches,

M

hyperbolic paraboloids, gable ties to barrel vault and folded plate roofs, suspenders to tied arch bridges and ties beneath prestressing beds. Extensions of ties are often desired to be as small as possible. This means a low strain is desirable in a tie, hence a steel tie or a prestressed concrete tie is designed, using a low stress. If the steel tie needs to be clad to resist fire or corrosion then the prestressed tie is often a more economical solution. One objection to steel ties to concrete structures is that their life and fire resistance is far less than that of the concrete members and if they failed a heavy structure would collapse. Pretensioning was favoured because the long slender members were considered to buckle as Euler's theory when post-tensioned. Refs. 10 and 11 show that the tendons restrain such buckling and a position of static equilibrium can be obtained when post-tensioning, so that if a certain unnoticeable curvature is allowed then the post-tensioned member can be designed accordingly and is a very economic design. Post-tensioned ties have the economic advantage that they can easily be effected on site from existing scaffolding to shells, arched bridges, etc., when required. Pretensioned ties have to be delivered on time and threaded amongst the scaffolding and provided with special end attachments. Designs of pretensioned and steel ties are compared in Ref. 12 and post-tensioned ties are designed in Refs. 10 and 11.

8.4.11 Shear resistance of prestressed concrete beams

At working loads for CP 110 Class 1 and 2 structures, beams are considered as uncracked and hence principal stresses can be calculated in the usual manner by combining stresses due to prestressing, bending and shear. The concrete is usually well able to resist the principal compressive stresses, and can usually resist the principal tensile stresses; if it cannot, then the section or the amount of prestressing has to be altered, or shear reinforcement in the form of inclined tendons, or vertical or inclined stirrups, or vertical prestressing, has to be introduced. The principal stresses can be calculated from the well known expression

$$f = 0.5\{f_h + f_v \pm \sqrt{[(f_h - f_v)^2 + 4v^2]}\} \tag{8.7}$$

where f_h and f_v are horizontal and vertical direct stresses (tensile positive) and $v =$ shear stress.

In the early days of prestressed concrete it was only necessary to limit the principal tensile stress to zero or a small amount, say 0.5 N/mm^2 at working loads. This can still be done for a preliminary design. Research in shear generally shows that, with the kind of load factors used, if a beam is satisfactory with regard to its ultimate shear resistance then the diagonal cracks at working loads for reinforced concrete beams are adequately narrow and they are narrower still for prestressed concrete beams because of the prestressing forces tending to close such cracks. CP 110 therefore regulates only the ultimate shear resistance and equation 8.7 is no longer useful as it was with CP 114. Research concerning ultimate shear strength[13, 14], as with non-prestressed concrete, is inconclusive and appears inconsistent, hence empirical formulae have to be agreed for codes and these have to err greatly on the side of safety in some instances, because of the erratic nature and sensitivity to many variables of shear failures.

Example 8.12. Consider the CP 110 design of the beam of Example 8.7 in shear.

At support where section is not cracked in bending and shear force is a maximum, suppose $f_{cp} = 7$ N/mm², then from Table 39, $V_{co} = 2.2 \times 250 \times 450$ N $= 247.5$ kN. Then CP 110 is concerned about sections where there is likely to be a bending crack towards failure reducing the shear strength of the beam. The C. and C.A. *Handbook* on CP 110 suggests considering such a crack at a distance of half the effective depth from the point of maximum bending moment. Suppose for such a section $V/M = 0.032$ m⁻¹ and $f_{pt} = 11$ N/mm². Now $I = 0.25 \times 0.45^3/12 = 0.001\ 898$ m⁴, hence $M_0 = 0.8 \times 11\ 000 \times 0.001\ 898/(0.375 - 0.1923) = 91.42$ kN m. For Table 5, $100A_s/bd = 100 \times 2 \times 491/(250 \times 375) = 1.047$, thus $v_c = 0.75 + 0.2 \times 0.047 = 0.759$ N/mm². From Example 8.6, $f_{pe} = 570$ N/mm² and $f_{pu} = 500/491 = 1.018$ kN/mm². Thus from CP 110, equation 46:

$$V_{cr} = \left(1 - \frac{0.55 \times 570}{1018}\right) \times 759 \times 0.25 \times 0.375 + 91.42 \times 0.032 = 52.17 \text{ kN}$$

These values of V_{co} and V_{cr} are required to be less than the shear forces due to ultimate loads at these sections.

8.4.12 Inclined tendons

If a tendon is inclined upwards, at an angle α to the horizontal, towards and to the support, it is easy to imagine in the case of post-tensioning with a force P that reactions P and R_1 are imposed on

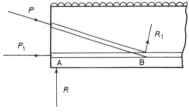

Fig. 8.7

the concrete (see *Figure 8.7*); thus for any section between A and B the shear force will be that due to the loading minus the vertical component of *P*, namely *P* sin α. Analyses giving shear forces and bending moments for beams with cables displaced upwards towards the supports with various profiles are given in Ref. 9.

8.4.13 Composite construction

An example of this is shown in *Figure 8.8*. The prestressed rectangular beam is propped until the *in situ* reinforced concrete slab is mature.

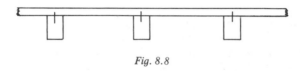

Fig. 8.8

Figure 8.9(a) shows the final T-beam. *Figure 8.9(b)* gives the stress distribution in the prestressed concrete beam before the slab is cast. There are dowel bars between the beam and the slab so that, when the props are removed, the self-weight of the slab and future live loading is carried by the 'composite' T-beam. *Figure 8.9(c)* shows the stress distribution after the props have been removed, the beam having to carry the self-weight of the slab, and *Figure 8.9(d)* shows the stress

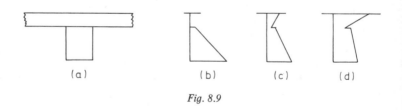

(a)   (b)   (c)   (d)

Fig. 8.9

distribution when the live loading is also being carried. The dowels required can be calculated by determining the horizontal shear stress at the junction between slab and rectangular beam (see Section 3.3). Composite construction is generally economic when a floor or bridge deck is desired to be *in situ* as opposed to precast and its total depth is required to be less than for *in situ* reinforced concrete construction or when durability (absence of cracks at working loads) is required (e.g. bridge decks).

8.4.14 Continuity

This has problems in that for various combinations of live loads on different continuous spans the tendons ideally need to be in varying positions. Cables have to be waived over supports of continuous beams; this increases friction losses and can make grouting difficult. Many calculations of sequence of prestressing and different loading possibilities have to be made. The careful control of the sequence of prestressing makes this operation more costly. A continuous beam shortens due to prestressing, so if the columns supporting it are *in situ*, ideally all but one need to be hinged at top and bottom so that some of the post-tensioning is not absorbed in bending the columns as opposed to post-tensioning the beam. A continuous beam is very vulnerable to the slightest differential settlement of supports. They can be designed for some settlement of supports and this makes them less economic. This is done for cable stayed prestressed concrete bridges used in Germany. Continuity is unpopular in the U.K. and is not favoured in mining subsidence areas; the jacking of supports requires too much attention and one could be caught out by sudden unpredictable settlement. Continuity is not to be recommended in the author's opinion, unless essential.

8.4.15 End splitting forces

Referring to *Figure 2.9*, the prestressing wire upon release increases its diameter at A, and thus splitting forces are created between, and normal to a line between, A and B. Designers have sometimes been unaware of this problem and have experienced splitting cracks in pretensioned members along the line between A and B. Other end splitting forces are caused by the end anchorages of tendons being, in effect, a system of irregularly distributed point loads on the end of a member. Each point load causes splitting forces normal to its line of action. Again failures have occurred.

This problem should be considered by the designer and CP 110 gives simple empirical guidance.

8.4.16 Prestressed concrete tanks, pipes, domes, shells and piles

For circular tanks and pressure pipes, circular prestressing is provided to counteract the circumferential tension due to the loading. A residual circumferential compression can ensure no cracks developing, due to shrinkage and temperature change, and this increases the watertightness. The pipes also need longitudinal prestressing for handling purposes. The writer has been consulted concerning troubles with certain prestressed concrete pipes. From his considerable literature searches he would recommend Ref. 15 for determining soil pressures on pipes and Ref. 16 for guidance on the design of prestressed concrete pipes. Some recent research supervised by the author on this problem is given in Ref. 17.

With rectangular tanks the walls must be free at the base, otherwise the corners act rigidly as folded plates and prevent the post-tensioning imposing stresses along the walls; a very able designer overlooked this point. Prestressing is useful for providing the ring tension to domes. The writer has rectified a dome, failing due to inadequate ring steel, by prestressing around the periphery.

Prestressed concrete piles are used for longer piles when handling stresses are a problem; end reinforcement details are important. Prestressing is useful for normal and North Light barrel vault roofs longer than about 36 m and 27 m respectively, assuming width-to-length is about 1:2.

8.4.17 Torsional resistance

The ultimate limit state for torsion is dealt with in the same way as for non-prestressed beams (see Chapter 3).

REFERENCES

1. Walz, K., 'Requirements concerning the grout for prestressed concrete elements', *Bau und Bauindustrie* (Düsseldorf), **8**, 16, p. 468 (1957) (in German)
2. Leonhardt, F., 'On the injection of the grout into the ducts', *Proc. Third Congress of the Fédération Internationale de la Précontrainte* (Berlin, 1958), Cement and Concrete Association, pp. 323–336 (1958) (in German)
3. Leonhardt, F., *Prestressed Concrete for the Practice*, Wilhelm Ernst und Sohn, 2nd Ed., p. 45, Berlin (1966) (in German)
4. Benz, G. H., *Grout for the Ducts of Prestressed Concrete*, Sittler and Federman, Illertissen, Germany (1965) (in German)
5. Soroka, I., and Geddes, J. D., 'Cement grouts and the grouting of post-tensioned prestressed concrete', *Bulletin No. 26*, University of Newcastle-upon-Tyne, Dept. of Civil Engineering, June (1966)
6. Szilard, R., 'A survey of the art—corrosion and corrosion protection of tendons in prestressed concrete bridges', *Journ. Amer. Conc. Inst.*, Jan. (1969)
7. Wilby, C. B., 'Precast concrete framed roofs—design of joints and use of post-tensioning', *Indian Concrete Journal*, Feb. (1960)
8. Winter, G., and Nilson, A. H., *Design of Concrete Structures*, McGraw-Hill, U.S.A. (1973)
9. Wilby, C. B., *Prestressed Concrete Beams*, Elsevier Publishing, Amsterdam,

London, New York; Applied Science Publishers, London (1969)
10. Wilby, C. B., 'Design of post-tensioned prestressed concrete ties for bridges and shell roofs', *Civil Engineering and Public Works Rev.*, Apr. (1972)
11. Wilby, C. B., *Elastic Stability of Post-tensioned Prestressed Concrete Members*, Ed. Arnold (1964)
12. Evans, R. H., and Wilby, C. B., *Concrete—Plain, Reinforced, Prestressed and Shell*, Ed. Arnold (1963)
13. Wilby, C. B., and Nazir, C. P., 'Shear strength of uniformly loaded prestressed concrete beams', *Civil Engineering and Public Works Rev.*, Apr. (1964)
14. Wilby, C. B., and Inman, P., 'The structural behaviour of jointed prestressed concrete beams,' *Proc. I.C.E.*, Sept. (1972)
15. Clarke, N. W. B., *Buried Pipelines: a Manual of Structural Design and Installation*, Maclaren (1968)
16. Swanson, H. V., 'Design of prestressed concrete pressure pipe', *Journal Prestressed Concrete Institute*, Chicago, Illinois, **10**, Aug. (1965)
17. Johnson, P., *Structural Design of Buried Pipes*, MSc Thesis, University of Bradford (1973)

Chapter 9

Shell and folded plate roofs

9.1 Shell roofs

9.1.1 *Economics of cylindrical shell roofs in the U.K.*

When the author was designing shells in the early days of popularity of shells in the U.K., about 1953, good designs were very arduous to make, it was very difficult for designers to learn how satisfactorily to design shells, the curved shuttering was very expensive, and as the contractors and their artisans were not experienced in this type of construction, shell roofs tended to be very expensive. Shell roofs enjoyed their popularity at that time chiefly because certain ones were constructed and publicised, many architects liked their aesthetic possibilities and some architects and engineers were interested in being involved and in obtaining experience in a new type of construction.

Subsequently designs can be made with much more confidence, as experience has enabled certain practical problems to be conquered (for example, where reinforcement is necessary to prevent cracking irrespective of but in addition to the complex elastic design analysis) and with the advent of computers[1, 2, 3] the designs are much less laborious. Designs are thus less expensive than they were, relative to designs of non-curved structures. With the development of special systems of curved shuttering the author found on one occasion that subcontracting the curved shuttering to a specialist firm was slightly less expensive than the estimated cost of the flat shuttering by the contractor's own labour. Many contractors now have personnel familiar with constructing shells. Hence shells are reasonably economic in the U.K., certainly compared with roofs of similar quality. However, there is a large range of prices and qualities of roofs one can obtain; for example a shell roof with good insulation, three layers of built-up roofing felt (the top one mineral finished), and plastered and emulsion painted beneath, can be about three times the price of a roof sheeted with corrugated asbestos with an underlining board to give thermal insulation supported by steel purlins and trusses, or by precast concrete purlins and pitched portal frames. It could be considered, particularly if the roof covered expensive machinery or an area of quality, that the shells are worth more than three times the price of the other roofs mentioned from a durability and water-tightness point of view. In addition, better aesthetics are possible with the shells than with the corrugated asbestos sheeting. In concrete one has a structure with low maintenance costs and good resistance to fire compared with similar steelwork roofs. If one requires these advantages and hence uses reinforced or prestressed concrete, then in these media the shells enable large spans to be covered economically and without large deflections. Folded plates provide a similar solution but require more materials and are hence heavier than curved shells and then do not carry these heavier loads as efficiently. However, it is surprising that more folded plates have not been used, particularly in the early days of shells in the U.K., when cylindrical shells were so much more difficult to design.

Today it undoubtedly pays the normal office to employ an expert, and the author has been employed in this capacity by consultants, designers and architects, to design shells; yet someone in the office could perhaps produce a reasonable folded plate design, although some of the experience of shells is also useful in the design of folded plates to be more crack-free and less troublesome. This advantage of folded plates—more easily tackled by the average office—should have done more than it has to assist the use of more folded plates, particularly as folded plates can have quite good aesthetic properties. There seems to have been, unfortunately and too often, a tendency for designers, instead of engaging experts to help them with the design of shell schemes requested by architects, to pose all the possible disadvantages, exaggerating high cost for example, to alter the scheme to something much easier for their office to design. It is felt by some that this has spoilt the development in use of shells in the U.K. since the initial enthusiastic period. To be fair, many of the early shells were excessively costly because of unfamiliarity, as outlined previously, with this new form of construction, and they acquired a reputation in many quarters for being expensive. However, although many buildings need to be inexpensive, many do not, hence the latter point is perhaps not as important as the former.

One would hope there would be a future intelligent use of shells and folded plates, obtaining the benefits of these structures where appropriate, giving better buildings aesthetically than the too familiar box-like buildings, irrespective of the design difficulties outlined, which have been reduced considerably with the use of computers[1,2,3].

9.1.2 *Proportioning of cylindrical shells*

Many cylindrical shells have now been designed and constructed in the U.K. and this experience enables reasonably accurate assessments to be made of the dimensions of many such shells for estimate purposes. The proportions to be suggested will suit many practical cases and will usually be found to be satisfactory upon analysis.

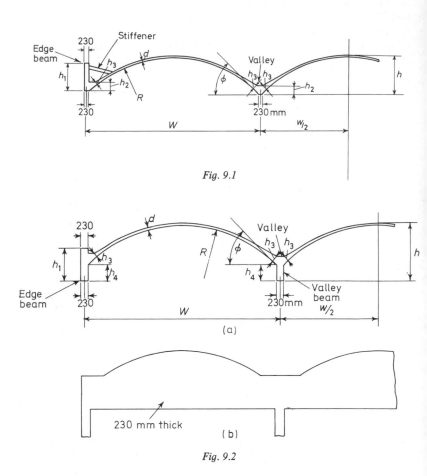

Fig. 9.1

(a)

(b)

Fig. 9.2

Figures 9.1 and 9.2(*a*) illustrate typical cross sections of normal barrel vault roofs (the shells are symmetrical apart from the end spans), and *Figure* 9.3(*a*) shows a typical cross section of a system of North Light barrel vault roofs. The following practical points refer to normal barrel vault roofs. Using the notation shown in these illustrations, if W is large compared with L (the span between end stiffening beams) then the structure chiefly comprises stiffening beams,

175

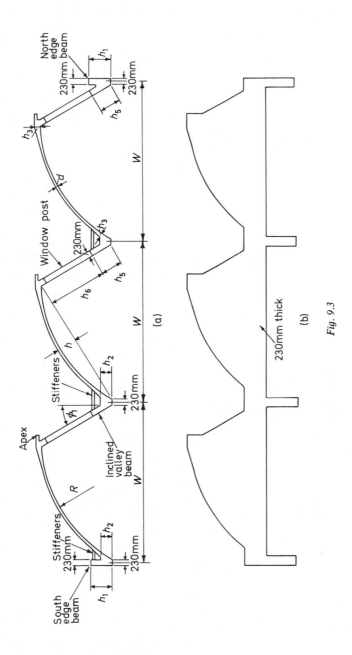

Fig. 9.3

little use being made of the shell action of the slab. Sometimes in these instances R is taken as equal to W. In the extreme case, the shell slab can be designed ignoring its curvature, without a great loss in economy. Short shells spanning between large arches (in lieu of stiffening beams) are extremely useful for spanning very large distances, e.g. several aircraft hangars have been constructed in this manner in the U.S.A., one such scheme having a clear span of 113 m ($= W$ in this instance). If W is small compared with L then a relatively large proportion of the area covered comprises valleys, together with the necessary thickenings and usually valley beams in these instances, as in *Figure 9.2*. Otherwise the shell at the springings in a system such as in *Figure 9.1* becomes very steep, increasing construction costs; hence full advantage is not being taken of the economy of materials associated with the thin shell portion of the construction. Such constructions are sometimes necessary for large spans, especially if loads are hung from the shells or the valley beams.

The author[4] designed and constructed some shells in 1957 which had to allow for modest future loads to be hung from anywhere, as could be done with the competitive steelwork. Steel tubes at 914 mm centres were cast in the valley beams and cadmium plated steel bolts were cast to protrude from the shell at 914 mm centres in both directions. In this connection the maximum practical span L for normal barrel vault roofs is approximately 37 m, if the shell is not prestressed or arched longitudinally; in the latter cases the maximum practical span is approximately 61 m. In the U.K. certain specialists consider that the most suitable ratio of L to W for economy is approximately 2:1, using the system shown in *Figure 9.1*. If h is low compared with L then insufficient shell action is obtained for economy and if h is high compared with L then too much material may be used and the springings of the shell can become too steep for economy. The optimum ratio of h to L for economy consistent with strength and deflection is considered to be approximately 1:10; sometimes 1:12 is used for spans less than about 18 m. If the edge beams are supported by intermediate columns then h_1 may be a nominal amount and 760 mm has been found to be satisfactory from experience—in this case *stiffeners* are not usually required. For unsupported edge beams a ratio of h_1 to L of 1:15 is usually satisfactory, and stiffeners are usually required; these are sometimes of the order of 150 mm square in cross section at third points of the spans, but have to be decided from the accurate design to prevent lateral instability of the compression zone of the edge beam. The angle ϕ should not exceed 45° if double shuttering and other extra expenses associated with steep springings to shells are to be avoided. If the above suggested proportions of h, L and W are used, ϕ will automatically be less than 45°. Shrinkage and temperature movement are not normally considered in the analysis of barrel vault roofs, and early experience showed that shells needed thickening near their springings to avoid cracking due to these effects. Hence h_2 should be not less than 460 mm and h_3 can be taken as $L/120$, with upper and lower limits of 230 mm and 130 mm. A uniform change of temperature should not alter stresses if the shell is unrestrained and efforts are made to do this by placing soft materials allowing small movements between the shells and any surrounding structures, and indeed the brickwork cladding to the shell structure. An alternative with brick-work is to use lime (not cement) mortar, which allows considerable movement without apparent cracking and fracturing of brickwork. Shrinkage is resisted by reinforcement and the concrete may crack to some extent, so that there is little total structural movement due to shrinkage compared to what one might expect from the shrinkage coefficient. Longitudinal *spacer* bars are provided in the shell for this purpose, as well as supporting the top steel and assisting concrete placing. The thickness of the shell d is usually not less than 65 mm in the U.K. and when W exceeds 13.7 m d is often increased to 75 mm; when W exceeds 18.3 m, d is often increased to 90 mm. Too great a thickness of shell is uneconomical, because the self-weight of the shell is the largest proportion of the load to be carried, e.g. a 65 mm shell weighs 1.56 kN/m^2, out of a total design load of say 2.4 kN/m^2. It is usual in the U.K. to increase the thickness d by 25 mm for a dis-tance $L/10$ from each end. This is necessary, together with extra reinforcement, to resist moments ignored by many analyses (e.g. those due to Schorer and Finsterwalder) but yet which are not negli-gible near to the supports. If barrel vault roofs need to be narrower than suggested previously, i.e. $W < 0.5\,L$, then the system shown in *Figure 9.1* results in the shell being too steep near the springings. Consequently the system shown in *Figure 9.2* is preferable in such instances, and h_4 is made so that ϕ is less than 45°. Sometimes valley beams are necessary for supporting cranes or suspended loads.

The following practical points refer to a North-Light (asym-metrical) system as shown in *Figure 9.3*. Again $W = 0.5L$ gives an economical solution. The curvature of the shell is often reduced as much as possible so that it is not too steep at the valley, and also for architectural appearance; h can be taken as $L/18$ but should not normally be reduced much below this amount. The value of h_1 is as before or sometimes deeper to suit the requirements of the gutter. In the U.K. ϕ_1 is often taken as 30° and in this instance $0.075L$ is usually a satisfactory value for h_5 and $0.22L$ for h_6; d, h_2 and h_3 are as before.

Stiffening (or gable) beams, arches and trusses are frequently made

230 mm wide, except when they are upstand, when they need to be 380–460 mm wide according to the design. Fairly common stiffening beams are shown in *Figures 9.2(b)* and *9.3(b)*. Columns can quickly be assessed for estimate purposes by considering conservatively only vertical load—unless cranes or other loads are supported in addition —and are not usually made less than 300 mm square because of the difficulty of detailing reinforcement at junctions with stiffening and valley beams.

9.1.3 *Estimating reinforcement for cylindrical shells*

The methods of Section 9.1.2 enable the dimensions of a preliminary scheme to be determined, and the amount of reinforcement required for such a scheme can be estimated from *Figure 9.4* accurately enough

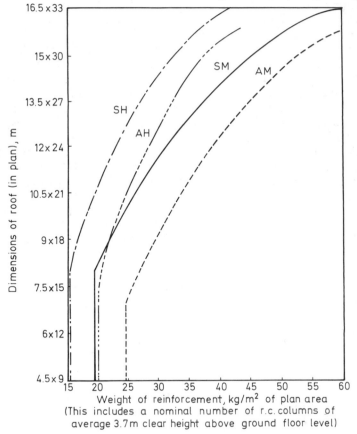

Fig. 9.4

for estimating the total contract price. Curves SH and AH are for normal and North Light shells respectively, where high tensile reinforcement is used apart from mild steel for ell-bars[5], stirrups and column bars. Curves SM and AM are for normal and North Light shells respectively, where mild steel is used throughout except for shell fabrics and their spacer bars[5] (Section 2.4). Designer suppliers of reinforcement would probably require more precise methods[5].

9.1.4 Design of shells

The design consists of estimating the dimensions of the members and then analysing elastically for the internal stresses. Reinforcement is designed to resist the tensile stresses, and the permissible concrete compressive stresses must not be exceeded; deflections can also be calculated. If the reinforcement is too heavy to fit into the sections, if the concrete compressive stresses or the deflections are too great, then the dimensions of the shell have to be altered and the analysis repeated until a satisfactory solution is found. Experience and approximate designs (such as the beam analogy)[6] can usually decide upon dimensions which will not need to be altered as a result of the subsequent elastic analysis. The information in Section 9.1.2 is based on experience and can be used in this way.

9.1.5 Elastic analysis of cylindrical shells

Elastic analyses are very arduous in practice (including even the part slide rule, part table method of Tottenham[7, 8]). Computer methods are thus highly desirable. The electronic analogue computer has certain advantages over the electronic digital computer in that it allows a semi-design technique, is much faster and can speedily graph the results on an inexpensive $X-Y$ plotter. Originally the analysis had to be split into statically determinate (membrane) and statically indeterminate (edge load) cases respectively to effect a solution. This tends to make solutions complicated for the beginner to follow because of the care needed with organisation of equations and signs. (Ref. 5 gives a simple explanation of the traditional analysis and gives equations which can easily be solved with a digital computer.) The analogue computer seemed useless for solving the analysis in this traditional way, as equating the displacements of the shell edge with its point of contact on the edge beam demands accuracy of the order of 7 to 10 significant figures with the traditional analysis whereas the analogue computer has only approximately slide rule accuracy, which is of course adequate for the resultant forces and moments per unit length. The author had to reunite the analysis and use a relaxation type of process, which is very rapid, to enable solution with the analogue computer[1,2]. When all this has been achieved, the author finds the basic analysis required more straightforward to teach to students. Hence this analysis is now developed.

For the element shown in *Figure 9.5(b)* of the shell shown in

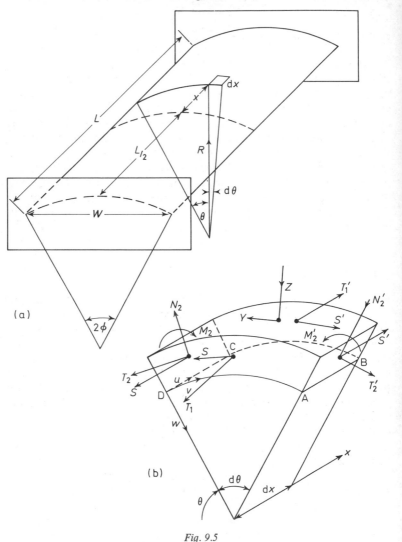

Fig. 9.5

Figure 9.5(a) (note that all forces and moments such as N_2 and M_2 are per unit length) taking moments about AB

$$M'_2 \, dx - M_2 \, dx - N_2 \, dx \, R \, d\theta + Z \, dx \, R \, d\theta (R \, d\theta/2) = 0$$

$$\therefore [M_2 + (\partial M_2/\partial\theta) \, d\theta] - M_2 - N_2 R \, d\theta + (ZR^2/2)(d\theta)^2 = 0$$

$$\therefore N_2 = (1/R)(\partial M_2/\partial\theta) \tag{9.1}$$

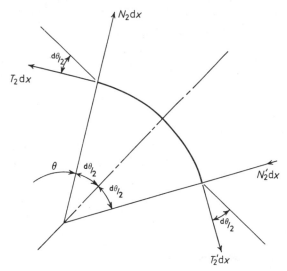

Fig. 9.6

Resolving normal to surface of element (see *Figure 9.6*)

$$T_2 \, dx \sin (d\theta/2) + T'_2 \, dx \sin (d\theta/2) - N_2 \, dx \cos (d\theta/2)$$
$$+ N'_2 \, dx \cos (d\theta/2) + Z \, dx \, R \, d\theta = 0$$

$$(T_2/2) \, d\theta + [T_2 + (\partial T_2/\partial\theta) \, d\theta] (d\theta/2) - N_2$$
$$+ [N_2 + (\partial N_2/\partial\theta) \, d\theta] + ZR \, d\theta = 0$$

$$\therefore T_2 + (\partial T_2/\partial\theta)(d\theta/2) + (\partial N_2/\partial\theta) + ZR = 0$$

$$\therefore T_2 = - (\partial N_2/\partial\theta) - ZR \tag{9.2}$$

Resolving tangentially to surface of element (see *Figure 9.6*)

$$N_2 \, dx \sin (d\theta/2) + N'_2 \, dx \sin (d\theta/2) + T_2 \, dx \cos (d\theta/2)$$
$$- T'_2 \, dx \cos (d\theta/2) + SR \, d\theta$$
$$- S'R \, d\theta + Y \, dx \, R \, d\theta = 0$$

N

$$\therefore N_2(d\theta/2) + [N_2 + (\partial N_2/\partial\theta)\, d\theta]\,(d\theta/2) + T_2$$
$$- [T_2 + (\partial T_2/\partial\theta)\, d\theta] + SR(d\theta/dx)$$
$$- [S + (\partial S/\partial x)\, dx]\, R(d\theta/dx) + YR\, d\theta = 0$$
$$\therefore N_2 + (\partial N_2/\partial\theta)(d\theta/2) - (\partial T_2/\partial\theta) - R(\partial S/\partial x) + YR = 0$$
$$\therefore \partial S/\partial x = -(1/R)(\partial T_2/\partial\theta) + (N_2/R) + Y$$

Schorer[9] and Jenkins[10] (but not Wilby[2]) assume that N_2/R is very small for thin shells and can be neglected.

$$\therefore \partial S/\partial x = -(1/R)(\partial T_2/\partial\theta) + Y \qquad (9.3)$$

Resolving longitudinally for element (refer to *Figure 9.5(b)*)

$$T_1 R\, d\theta - T_1' R\, d\theta + S\, dx - S'\, dx = 0$$
$$T_1 R - [T_1 + (\partial T_1/\partial x)\, dx]\, R + S(dx/d\theta)$$
$$- [S + (\partial S/\partial\theta)\, d\theta]\,(dx/d\theta) = 0$$
$$\therefore -R(\partial T_1/\partial x)\, dx - (\partial S/\partial\theta)\, dx = 0$$
$$\therefore (\partial T_1/\partial x) = -(1/R)\,(\partial S/\partial\theta)$$
$$\therefore (\partial^2 T_1/\partial x^2) = -(1/R)\,(\partial/\partial\theta)(\partial S/\partial x) \qquad (9.4)$$

Considering longitudinal strain, the element of length dx has a displacement of u at $x = x$ and of $u + (\partial u/\partial x)\, dx$ at $x = x + dx$ (see

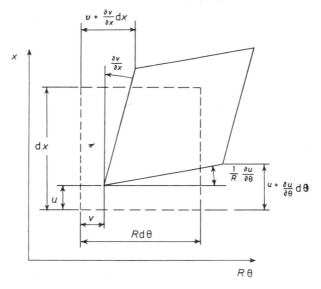

Fig. 9.7

Figure 9.7). The change in length is therefore $(\partial u/\partial x)\,dx$ and the strain therefore $\partial u/\partial x$. The corresponding stress is T_1/d (see *Figure 9.5(b)*). Thus applying Hooke's law $T_1/d = E(\partial u/\partial x)$ where E is Young's modulus

$$\therefore (\partial u/\partial x) = T_1/(Ed) \qquad (9.5)$$

Considering shear strain of element (see *Figure 9.7*)

$$\text{Shear strain} = \omega = (1/R)\,(\partial u/\partial \theta) + (\partial v/\partial x)$$

$$\therefore (\partial v/\partial x) = -(1/R)\,(\partial u/\partial \theta) + \omega$$

It is reasonable to assume that ω is much less than $(1/R)\,(\partial u/\partial \theta)$

$$\therefore (\partial v/\partial x) = -(1/R)\,(\partial u/\partial \theta)$$

$$\therefore (\partial^2 v/\partial x^2) = -(1/R)\,(\partial/\partial x)\,(\partial u/\partial \theta) = -(1/R)\,(\partial/\partial \theta)\,(\partial u/\partial x)$$

Substituting for $\partial u/\partial x$ from equation 9.5 we obtain

$$(\partial^2 v/\partial x^2) = -[1/(REd)]\,[\partial T_1/\partial \theta] \qquad (9.6)$$

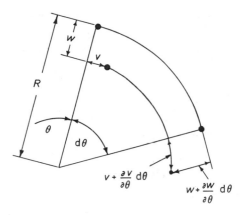

Fig. 9.8

Consider that in *Figure 9.8*, the element, to move to its stressed position, first moves radially and then moves circumferentially; the tangential strain due to each of the movements is:

1. Owing to radial movement of the element its length is reduced by

$$R\,d\theta - (R - w)\,d\theta = w\,d\theta$$

2. Owing to the circumferential movement of the element its length is increased by $(\partial v/\partial \theta)\,d\theta$.

Thus the change in length of the element $= (\partial v/\partial\theta)\,d\theta - w\,d\theta$. This gives a tangential strain of

$$[(\partial v/\partial\theta)\,d\theta - w\,d\theta]\,[1/(R\,d\theta)] = (1/R)\,[(\partial v/\partial\theta) - w]$$

which can also be evaluated as T_2/Ed (see *Figure 9.5(b)*)

$$\therefore\ (\partial v/\partial\theta) - w = RT_2/(Ed)$$

$$w = -[RT_2/(Ed)] + (\partial v/\partial\theta) \tag{9.7}$$

It is reasonable to assume that $RT_2/(Ed)$ is much less that $\partial v/\partial\theta$

$$\therefore\ w = \partial v/\partial\theta \tag{9.8}$$

Again considering *Figure 9.8*, the change in slope of the shell

$$= \psi_2 = \frac{(\partial w/\partial\theta)\,d\theta}{R\,d\theta} = (1/R)\,(\partial w/\partial\theta) \tag{9.9}$$

Then from simple beam theory $[M/(EI) = -(d^2 y_1/dx_1^2) = -(d/dx_1)(dy_1/dx_1) = -(d/dx_1)\psi_2 = -d\psi_2/(R\,d\theta)]$

$$M_2/(EI) = -(1/R)\,(\partial\psi_2/\partial\theta) \tag{9.10}$$

By using load harmonics (i.e. Fourier's series) each variable symmetrical in x may be expressed in terms of its values at the centre ($x = 0$) thus: $f(x) = f(0).\cos ax$, where $a = (n\pi/L)$ ($n = 1, 3, 5$, etc.) and it has been found that it is reasonable to consider $n = 1$ only. Differentiating twice with respect to x

$$(\partial^2/\partial x^2)\,f(x) = -a^2 f(0).\cos ax = -a^2 f(x)$$

Hence in such instances $\partial^2/\partial x^2$ can be replaced by $-a^2$. Hence equation 9.4 becomes

$$T_1 = [1/(a^2 R)]\,(\partial/\partial\theta)\,(\partial S/\partial x) \tag{9.11}$$

and equation 9.6 becomes

$$v = [1/(a^2 REd)]\,(\partial T_1/\partial\theta) \tag{9.12}$$

Re-arranging the previous equations to obtain them in a form suitable for the electronic analogue computer we have:

From equation 9.1 $\quad M_2 = \int_0^\theta RN_2\,d\theta + M_{2i}$ $\qquad(9.13)$

From equation 9.2 $\quad N_2 = \int_0^\theta (-T_2 - ZR)\,d\theta + N_{2i}$ $\qquad(9.14)$

From equation 9.3 $\quad T_2 = \int_0^\theta [-R(dS/dx) + RY]\,d\theta + T_{2i}$ $\quad(9.15)$

From equation 9.11 $dS/dx = \int_0^\theta (a^2 RT_1)\,d\theta + (dS/dx_i)$ $\qquad(9.16)$

From equation 9.12 $T_1 = \int_0^\theta (a^2 REdv)\,d\theta + T_{1i}$ $\qquad(9.17)$

From equation 9.8 $v = \int_0^\theta w\,d\theta + v_i$ $\qquad(9.18)$

From equation 9.9 $\quad w = \int_0^\theta R\psi_2 \, d\theta + w_i$ $\qquad\qquad$ (9.19)

From equation 9.10 $\psi_2 = \int_0^\theta - [R/(EI)] M_2 d\theta + \psi_{2i}$ $\qquad$ (9.20)

where the suffix i denotes initial condition when $\theta = 0$.

The preceding eight equations are suitable for solution with an electronic analogue computer, which relates voltage and time. For example, if a voltage is expressed as a mathematical function of time, then an *integrator* will integrate this function with respect to time. In the preceding eight equations θ is taken as analogous to time and each of the parameters M_2, N_2, T_2, dS/dx, T_1, v, w and ψ_2 is taken as analogous to a voltage that is a function of time, i.e. θ. Each of these parameters in the previous equations, as far as the electronic analogue computer is concerned, is thus a voltage expressed as a simple integral of time, and each can thus be solved using an integrator.

The analogue computer requires initial conditions, i.e. for setting the initial condition potentiometers, hence each integration is done from $\theta = 0$ to θ and added to the initial value of the parameter under consideration, as can be seen in the equations.

The electronic analogue computer circuit diagram for the above equations is shown in Fig. 2 of Ref. 1. Ref. 1 then gives the same approach for solving Jenkins[10] (often called D.J.K.) type of analysis. Ref. 1 then gives examples of symmetrical and North Light shells, solving them, using the above approach, in accordance with the Schorer and Jenkins types of analyses respectively. Ref. 2 uses the same approach and is based on a theory that makes even less assumptions than the D.J.K. equation. It deals with symmetrical and North Light shells and gives an example of a North Light shell to show how this more exact analysis has little extra trouble even with the more difficult North Light shell.

Equations 9.13–9.20 are the same as equations 28–35 in Ref. 1.

Boundary conditions, e.g. the statical conditions which must exist at the boundaries between shell and edge beam, are given for symmetrical and North Light shells in Ref. 1. One problem is that the analogue computer needs initial conditions, i.e. at $\theta = 0$. Basically we are solving an eighth order differential equation. Ref. 1 shows that the preceding eight equations can be resolved into the eighth order differential equation 27. Hence a traditional solution has basically eight arbitrary constants of integration which can only be determined by knowing eight boundary conditions. Hence we need to know eight boundary conditions, ideally at $\theta = 0$. For the shells considered we usually know some at $\theta = 0$ and the others at the crown ($\theta = \phi$) for a symmetrical shell and at the apex for a North Light shell roof. The way we devised using the analogue computer[1] is to guess ones that one does not know at $\theta = 0$ and then to see the error in the ones we know at the crown or apex. Then readjust these guesses until all is satisfied, i.e. a relaxation procedure. As the computer has such fantastic speed the adjustments take little time.

This method has an advantage over the digital computer in that if, say, a stress or deflection is too great one can soon determine which combination of parameters have the greatest influence on this stress, again because of the almost instantaneous speed of the analogue computer. It will often be a combination which could not have been guessed for trial and error on a digital computer. Hence this is a design, as opposed to analysis, technique to a useful extent.

Some MSc and BSc students are taught electronic analogue computation by computing departments, and the foregoing, as well as Appendix 2 of Ref. 11, can form useful exercises and/or projects for students of civil and structural engineering.

9.1.6 Elastic analysis of cylindrical shells using tables

Tables[3] have been produced for analysing a large range of practical shells. These give solutions of ordinary shells of spans 20–100 ft (6096–30 480 mm) in increments of 2 ft (610 mm), and of North Light shells of spans of 20–90 ft (6096–27 430 mm) in increments of 2 ft (610 mm). These increments are sufficiently small to allow linear interpolation.

Previous tables known to the author are not generally suitable for interpolation and have therefore very limited practical value. In addition, some tables give some unsuitable shells. The tables of Ref. 3 also give an example of the design and reinforcement for a North Light system of shells, including the various edge beams and glazing posts.

Example 9.1. Design the shells shown in *Figure 9.9*. These are continuous in the longitudinal (60 ft) direction and we find that if we analyse as if simply supported for a span of 60 ft (18 290 mm) and modify the forces and moments as follows, we do not seem to be more than about 10 % in error, which is quite reasonable for practical purposes and considering the great saving in design effort.

Hence we analyse the shells using the tables as demonstrated (for a 60 ft simply supported span) in the example of Chapter 4 of Ref. 3. We allow for continuity in the 60 ft direction as follows:

1. The transverse forces and moments T_2, N_2, M_2, are unaffected by continuity.
2. The longitudinal forces T_1 determined in Chapter 4 of Ref. 3 are at mid span of the simply supported 60 ft span and can be taken as proportional to $0.125\,qL_1^2$ (i.e. the maximum mid span bending moment for a simply

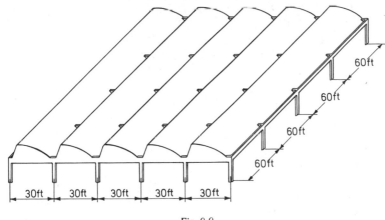

Fig. 9.9

supported span, $L_1 = 60$ ft, carrying a uniformly distributed load of q lb/ft). Then the T_1 forces at various sections are proportioned according to the bending moments at these sections for a continuous beam analogous to the continuous shells. In this example the bending moment coefficients for a beam of four equal spans are as shown in *Figure 9.10*. Thus the T_1 forces over supports B and D, for the shells, are the T_1 forces of the tables multiplied by 0.107/0.125. Similarly the T_1 forces at the section of maximum bending moment in the end spans are the T_1 forces of the tables multiplied by 0.077/0.125, *et cetera*.

3. The shear force S is proportioned using a continuous beam analogous to the continuous shells similarly to 2. In this example the shear force coefficients for a beam of four equal spans are as shown in *Figure 9.11*. For a simply supported span the shear force coefficient for sections immediately adjacent to the supports is 0.5. Thus the S forces immediately adjacent to B, in the span AB, for the shells, are the S forces of the tables multiplied by

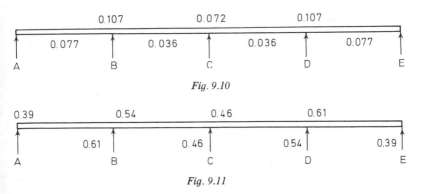

Fig. 9.10

Fig. 9.11

0.61/0.5. Similarly the *S* forces immediately adjacent to A are the *S* forces of the tables multiplied by 0.39/0.5, *et cetera*.

Continuous ordinary shells can be treated in the same way as for continuous North Light shells as previously, and can use the tables[3].

This approximate method is justified by research supervised by the author[12].

9.1.7 Elastic analysis of conoidal shells using tables

Tables[13] have been produced for analysing a large range of practical shells.

Example 9.2. Design the shells shown in *Figure 9.12*. *Figure 9.13* shows details of the cross section of the shells, and *Figure 9.14* a longitudinal cross section.

In *Figure 9.14* it can be seen that there is little continuity. It is reasonable to design these shells as though simply supported spans of 15 m. The end rotation is very small and adjacent shells sharing a transverse truss stiffener, provided the local reinforcement detailing is sensible, should be satisfactorily designed as just mentioned.

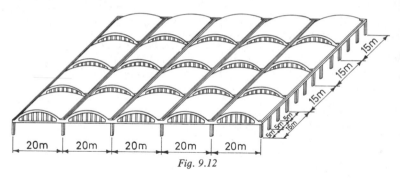

Fig. 9.12

From *Figure 9.13* it can be realised that all the internal shells are the same and can be designed from the tables[13], and this particular example is used in Chapter 4 (Section 4.3) of Ref. 13. With regard to the external shells of *Figure 9.13*, if there were no intermediate columns supporting the edge beam, the external shells would be designed just the same as the internal shells. In this particular example we have intermediate columns, so we design the outer half of each external shell as though it were half of a symmetrical shell with propped edge beams. This is done in Chapter 4 (Section 4.2) of Ref. 13. The inner other half of each external shell can be designed in the same way as the internal shells. After saying this, to be on the safe side for each external shell one would design the reinforcement for the shell portions of these two halves to be the same, designing the reinforcement in each location for whichever is the worst force or moment in the two halves. In other words, design the shell reinforcement to be symmetrical for each external shell. The forces and

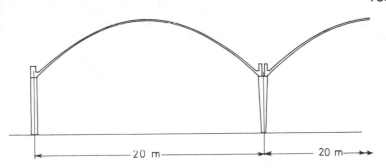

Transverse section of conoidal shell roof

(a)

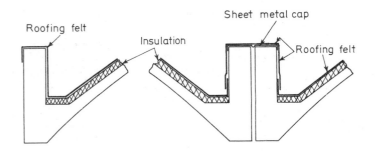

Enlarged sections of valley and edge beams

(b)

Fig. 9.13

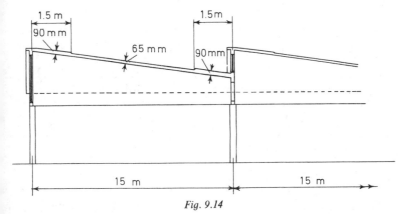

Fig. 9.14

moments in the two half-shells of each external shell influence one another because of their proximity. However, the edge beams of each external shell are remote enough to be considered not to influence one another. Hence for each external shell the supported external edge beam is designed as in Section 4.2 and the internal edge beam is designed as in Section 4.3 of Ref. 13.

To aid the reader in the use of tables[13] the following notes are useful:

Page 79, line 11, 1.321 should read 1.521.

Page 80, line 20, 0.5 should read 0.75.

Page 85, after the line 'Hence the flexural stresses in the extreme fibres are' it should read

$$'\pm (28.2 + 295) \times 10^6 \times 500/(1.94 \times 10^{10}) = \pm 8.33 \, \text{N/mm}^2$$

The direct stress $= 590\,140/(230 \times 1000) = 2.566 \, \text{N/mm}^2$

Height of neutral axis from soffit of beam $= (8.33 + 2.566)$
$$\times (500/8.33) = 654 \, \text{mm}$$

Maximum total tension in the edge beam
$$= 0.5 \times 230 \times 654 \times (8.33 + 2.566) = 819\,500 \, \text{N}$$
$\therefore$ number of 25 mm diameter bars needed
$$= 819\,500/(140 \times 490.9) = 11.92'$$

Page 86, line 1 should read 'Provide 12 number bars as shown in Figure 9'.

Figure 9, there should be 12 number 25 mm diameter bars instead of the 10 shown.

Page 83, Figure 7b, bars marked 1.01 should be alternately lifted and dropped so that they can be wired to the top and bottom steel respectively. The stirrups in the edge beams should be shown as complete rectangular stirrups. In the tables, except for V_y and the deflections, the columns relate to, from left to right, the crown, $\frac{1}{8}b$ from the crown, then $\frac{1}{4}b$, $\frac{3}{8}b$, $\frac{1}{2}b$, $\frac{5}{8}b$, $\frac{3}{4}b$ and $\frac{7}{8}b$ from the crown, then the shell edge. The rows relate to, from top to bottom, the high end, $\frac{1}{8}a$ from the high end, then $\frac{1}{4}a$, $\frac{3}{8}a$, $\frac{1}{2}a$, $\frac{5}{8}a$, $\frac{3}{4}a$ and $\frac{7}{8}a$ from the high end, then the low end. All this is obvious from the examples.

All edge beams have the dimensions given in Figures 3 and 7b. Vx and Vy given in the tables should be V_x and V_y as shown in the Notation. These are vertical support reactions per unit length (of curved length in the case of the transverse high and low ends). The two rows given in any table for V_x are firstly for the high end, $x = 0$, and secondly for the low end, $x = a$. The single row for V_y is for a longitudinal edge, $y = \pm b$. On pages 19–46 inclusive, the first row of the deflections, which are vertical, given on each page relates to the crown $(y = 0)$ and the second row is for $y = \pm(b/2)$. Hence for V_y and the deflections just mentioned the columns relate to, from left to right, the high end, $\frac{1}{8}a$ from the high end, then $\frac{1}{4}a, \frac{3}{8}a, \frac{1}{2}a, \frac{5}{8}a, \frac{3}{4}a$ and $\frac{7}{8}a$ from the high end, then the low end. On pages 49–76 inclusive, the first two rows of the deflections on each page relate to the same points as for pages 19–46. The third row relates to the vertical deflection of each edge beam.

9.2 Folded plates

The economic and aesthetic advantages of curved shell roofs, when employed over large unobstructed floor areas, also accrue to a considerable extent to *folded plate* roofs, also known as *hipped plate* or *prismatic* structures. They have additional advantages over curved shell roofs in that shuttering, steel fixing, setting-out, con-

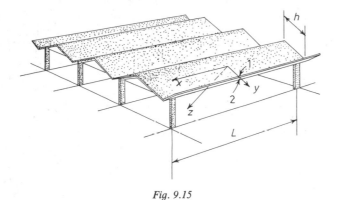

Fig. 9.15

creting, the carrying of point loads and incorporation of large openings are all much easier, but the weight and total quantities of concrete and reinforcement are greater. *Figure 9.15* shows a system of folded plates.

9.2.1 Design and analysis of folded plates

When they carry loads, internal forces and moments are induced in three dimensions as for cylindrical shells. Folded plates are designed by assessing their sizes by simple approximate design methods and/or experience, and then structurally analysing. The sizes are then altered if necessary and the process repeated. With experience it is often not necessary to alter the sizes and so in these cases the analysis gives confidence and allows economy of reinforcement compared to the original assessment.

9.2.2 Analysis of folded plates

A rigorous three-dimensional elastic analysis is given in Ref. 14. This has been programmed for digital computation[15]. Gibson[16] uses a program, based on the theory of Ref. 10 for analysing many cylindrical shells adjacent to one another, for analysing folded plates, making the curvatures of the shells very small. He defends this approximation of a shallow shell to a plate for a reasonable range of folded plates.

More approximate methods deal with less forces and moments. Resolving the loads in the directions y and z (see *Figure 9.15*) for each plate, then each plate can be considered as a beam of span L and depth h. If everything were symmetrical, that is if the folded plate system of *Figure 9.15* continued *ad infinitum*, without terminating at either end, then the longitudinal flexural stresses in these plates would be the same where they touch one another and the design just mentioned would be adequate, and can be called a *beam analogy theory*. Each plate is very flexible in the z direction but relatively very stiff in the y direction. Hence each apex point and valley point is held in position, in that the two adjacent plates restrict its movement in their planes (i.e. y directions). Hence away from the end columns the slab can be regarded as spanning from apex to valley as a long continuous slab, assuming the slab is not two-way spanning, i.e. $L > 2h$. Thus with the beam analogy method we obtain longitudinal stresses in the direction x and transverse bending moments and shear forces. For a large number of similar folded plates the beam analogy could be reasonable for the central bays.

For a scheme like *Figure 9.15* where there are end plates and not all the plates have the same dimension h, the beam analogy would mean that the longitudinal stresses calculated at say points 1 and 2 on *Figure 9.15* would not be the same. Yet they must be, as the longitudinal strain at these points must be the same. Thus Winter and Pei[17] introduce shearing forces along each fold and then determine these shear forces so that the longitudinal stresses at adjacent points (such as 1 and 2) at each fold (valleys and apices) are the same. These unknown longitudinal shears can be obtained by solving simultaneous linear equations by computer, calculating machine, or by a relaxation method (analogous to moment distribution) given by Winter and Pei. The transverse moments are calculated as before.

Gaafar[18] showed that in many cases, particularly at a place like the edge valley of *Figure 9.15* where two plates meet with very different values of h and they are discontinuous and continuous at their other ends respectively, the relative deflections of adjacent folds could make the Winter and Pei method inaccurate in assessment both of longitudinal stresses and of transverse moments. The relative deflections of the various folds (supports) have their effect upon the continuous slab (transverse) bending moments. This relative support settlement affects the reactions at the supports (folds) and thus the longitudinal stresses. This method is adequately accurate[19] for folded plates where $L > 3h$. In teaching students the presentation of this method by Simpson[20] is most useful, giving them a good insight into the principal behaviour of folded plates. Rapidly converging relaxation is used to eventually build up simultaneous equations, which essentially can be directly produced by Parme's method[21] by substituting numbers in equations (see Sections 9.2.3).

An analysis making very similar assumptions is given by Thadani[22] in terms of tabulated coefficients based on a method by W. Tetzlaff. Computer programs are given for this analysis by Tamhankar and Jain[23].

9.2.3 Analysis due to Parme

Parme applies the slope deflection equations to the transverse moments. Then the support reactions (at folds) are resolved in the planes of the plates, and the deflections of these plates in their planes are then equated to the unknown deflections used in the slope deflection equations. This gives linear simultaneous equations which can be solved by calculating machine or computer. Without this equipment they can be solved by relaxation. The following example, by courtesy of Dr A. L. L. Parme[24], is useful in indicating a procedure for a student or a practising engineer to pursue in his design office.

Table 9.1 in (a) shows the notation used for any type of folded plate. Then in (b) is shown the folded plate we have decided to analyse. This has been chosen to not look like the folded plate of (a), to indicate the general applicability of the method. When a plate is denoted by a number n, this number denotes the plate on its right-hand side. The length of the plates $= L = 70$ ft in this example, and W_n is determined from $W_n = h_n(w_D + w_L \cos \beta_n)$. In the example (density of reinforced concrete $= 150$ lb/ft^3) $n = 0$, self-weight concrete $= 0.5 \times 150 = 75$ lb/ft$^2 = w_D$, $W_n = 3 \times 75 = 225$ lb/ft
$n = 1$, self-weight concrete $= 0.333 \times 150 = 50$

Roofing and insulation $\qquad = \underline{5}$

$$w_D = \overline{55} \text{ lb/ft}^2$$

Table 9.1.

(a)

α positive when clockwise

(b)

Centre line

Given properties · *Trigonometric values*

(1) Point or plate	(2) α_n, deg	(3) β_n, deg	(4) t_{n-1}, ft	(5) t_n, ft	(6) h_{n-2}	(7) h_{n-1}	(8) h_n, ft	(9) h_{n+1}	(10) W_{n-2}	(11) W_{n-1}	(12) W_n, lb	(13) W_{n+1}	(14) $\cos\beta_{n-2}$	(15) $\cos\beta_{n-1}$	(16) $\cos\beta_n$	(17) $\cos\beta_{n+1}$	(18) $\sin\alpha_{n-1}$	(19) $\sin\alpha_n$	(20) $\sin\alpha_{n+1}$	(21) $\cot\alpha_{n-1}$	(22) $\cot\alpha_n$	(23) $\cot\alpha_{n+1}$
0		90		0.5000			3.000	10.01			225	800										
1	60	30	0.5000	0.3333		3.000	10.01	10.07		225	800	800			0	0.8660			0.8660			0.5774
2	20	10	0.3333	0.3333	3.000	10.01	10.07	10.07	225	800	800	800		0	0.8660	0.9848		0.8660	0.3420		0.5774	2.748
3	20	−10	0.3333	0.3333	10.01	10.07	10.07	10.01	800	800	800	800	0	0.8660	0.9848	0.9848	0.8660	0.3420	0.3420	0.5774	2.748	2.748
4	20	−30	0.3333		10.07	10.07	10.01		800	800	800		0.8660	0.9848	0.9848	0.8660	0.3420	0.3420	0.3420	2.748	2.748	2.748
5																						
6																						
7																						
8																						
9																						
10																						

Table 9.2.

Computed values

Point or plate	$\frac{⑧}{⑦}$ (24)	$\frac{⑲}{㉔}$ (25)	$㉒+㉓$ (26)	$㉔^2(㉑+㉒)$ (27)	$\dfrac{1}{24\times\left(\frac{⑤}{④}\right)^3}$ (28)	$\dfrac{⑤\times㉔}{6}\times\left(\frac{L}{8/\pi}\right)^2$ (29)	$\dfrac{⑤\times⑧}{6}$ (30)	$\left(\frac{⑤}{⑧}\right)^3=\left(\frac{7}{\pi}\right)^2$ (31)	$\dfrac{⑧\times㉔}{⑥\times⑱}$ (32)	$㉕+㉗+㉜$ (33)	$2\times㉕+㉖+㉗$ (34)	$㉕+㉖+㊱$ (35)	$\dfrac{⑨\times⑳}{⑧}$ (36)	$\dfrac{28}{24}$ (37)	$\dfrac{⑪\times⑮}{2}$ (38)	$\dfrac{⑫\times⑯}{2}$ (39)	$\dfrac{⑧\times㉚}{2}$ (40)	$\dfrac{㉔\times⑭(⑩+⑪)}{⑱}$ (41)	$(⑪+⑫)\left(\frac{⑮\times⑯+㉔}{⑲}\right)$ (42)	$\frac{⑰}{⑳}(⑫+⑬)$ (43)	$⑪\left(\frac{⑮\times⑯+㉔}{⑲}+㊷\right)$ (44)	$\dfrac{2\times㊳}{㉔}$ (45)	$\left[⑫\times\left(2+\frac{⑧\times⑯}{⑨\times⑰}\right)+⑬\right]\frac{⑰}{⑳}$ (46)
0																							
1	3.3337	3.853	3.325	3.365	0.4495	220.6		0.01799	3.900	10.21	14.75	10.09	0.3461	0.9881	0		330.9	0	3420	4607	4171	0	1250
2	1.006	2.942	5.496	5.496	0.9940	8.908	8.750	0.01799	2.942	11.36	16.84	11.36	2.907	1.000	346.4	393.9	44.58	0.4051	8686	4607			
3	1.000	2.924	5.496	5.496	1.000	8.750	8.750		2.942			11.36	2.924		393.9	393.9	44.06		9215	4051			
4													2.942				44.06						
5																							
6																							
7																							
8																							
9																							
10																							

Table 9.3.

Coefficients for simultaneous equations

① Point or plate	㊼ (30)×(36)	㊽ −(40)×(46)	㊾ (29)	㊿ 2((1)+(29))	(51) −(30)×(35)	(52) (30)×(36)	(53) (40)[−(43)+(44)+(43)((25)+(26))]	(54) −(31)×(33)	(55) (31)×(33)	(56) 2×(28)	(57) −(31)×(34)
0	76.35	−413 600									
1			0.4495	2.899	−89.88	25.90	−19 440				
2								−0.070 16	0.1837	1.988	−0.2654
3								−0.052 93	0.2044	2.000	−0.3030
4											
5											
6											
7											
8											
9											
10											

Table 9.3. (continued)

Coefficients for simultaneous equations

Point or plate (1)	4(1+(28)) (58)	(31)×(35) (59)	−(31)×(36) (60)	−(8)((37)×(38)+(39)) (61)	(30)×(32) (62)	(29) (63)	−(30)×(33) (64)	2(1+(29)) (65)	(30)×(34) (66)	−(30)×(35) (67)	(30)×(36) (68)	−(40)((41)−(42)+(43)) (69)
0												
1												
2	7.976	0.2044	−0.052 60	−7414	34.13	0.9940	−89.34	3.988	129.1	−99.40	25.58	179 700
3	8.000	0.2044	−0.052 93	−7933	25.74	1.000	−99.40	4.000	147.4	−99.40	25.74	49 040
4												
5												
6												
7												
8												
9												
10												

Table 9.4.

Point	Eq.	f_0	f_1	f_2	M_2	f_3	M_3	$= Load$
0	29	2	1		(47)			= (48)
1	27	(49)	(50)	1	(51)	(59)	(52)	= (53)
2	8	(54)	(55)	(57) + (60)	(58) + (68)	1	2	= (61)
	33		(63) + (60)	(65)	(66) + (56)	(57)	(67)	= (60)
3	8		(54)	(59) + (59)	(56) + 2	(65)	(58)	= (61)
	33			(63) + 1	(64) + (67)		(66)	= (69)

199

Table 9.5.

Point	Col.	0	1	2	3	4	5	6	7
	Row	f_0	f_1	f_2	M_2	f_3	M_3	= Load	Check column
0	0	2.000	1.000	0	76.35	0	0	= −413 600	−413 500
1	1	0.449 5	2.899	1.000	−89.88	0	25.90	= −19 440	−19 500
	2	−0.070 16	0.1837	−0.3180	7.976	0.2044	2.000	= −7414	−7404
2	3	0	0.9940	3.988	154.7	1.000	−99.40	= 179 700	179 800
	4	0	−0.1059	0.4088	4.000	−0.3030	8.000	= −7933	−7921
3	5	0	0	2.000	−198.8	4.000	147.4	= 49 040	48 990
Solutions		−169 800	−4554	33 890	−909.8	14 970	−1760		

Live load $= w_L = 25\,\text{lb/ft}^2$

$W_1 = 10.01\,(55 + 25 \times 0.866) = 767.3$

W_1, W_2, W_3 and W_4 are taken as 800 lb/ft in this example due to Parme.

In the U.K. w_L would be taken as 15 lb/ft^2 (snow load). The wind loading can be neglected as it is upwards and always less per unit area than the self-weight of each concrete plate. There is no reason why accurate values should not be taken for W_1, W_2, etc. *Tables 9.1, 9.2* and *9.3* are completed and it can be seen that for this example, as it is symmetrical, we do not need to consider all values of n. To illustrate how these tables are constructed, take for example the numbers in Column ㉟. They are obtained by adding the numbers in Columns ㉕, ㉖ and ㊱. The method of constructing the necessary simultaneous equations for solution of the problem is shown in *Table 9.4*. The equation numbers refer to mathematical equations in Parme's paper[21]. In all the tables so far given the shaded spaces eliminate quantities not required by the basic mathematical equations[21]. The numbers of the example are now inserted in *Table 9.4* to give *Table 9.5*. The equations (rows) are renumbered 0 to 5, for simplicity. These equations can easily be solved with a high accuracy electronic calculator, or with a standard program on a digital computer. With the former, the classical method of eliminating one unknown at once is better in practice than using determinants. Without such equipment an iteration method is recommended—a suitable one is demonstrated by Parme[21].

The solutions are given in the last row of *Table 9.5*. Note that all units used are lb ft units, i.e the stresses are in units of lb/ft^2. The symbols used are as follows:

M = Transverse bending moment at a fold and is considered positive when it creates tension on the underside of the plate (lb ft/ft length of plate)

f = Longitudinal stress and is considered positive when it is compressive (lb/ft^2)

W = Total vertical load acting on a plate, and is considered positive where acting downward (lb/ft length of plate)

These moments and stresses must be multiplied by $4/\pi$ to account for the difference in the sinusoidal load, assumed in the mathematical equations, and a uniformly distributed load. The use of a sinusoidal load is necessary for the mathematics and is a common device for many problems of elasticity. For example its use can be appreciated in Section 9.1.5 for cylindrical shells.

9.2.4 Design Tables for folded plates

With the assistance of Dr I. Khwaja the writer has prepared *Table 9.6* for designing certain useful symmetrical schemes of folded plates. A computer program similar to Ref. 16 was used. The folded plates are as shown in *Figure 9.16*, and $L = 20$ m, breadth of either edge beam $= 0.2$ m, thickness of each inclined plate $= 0.1$ m, Young's modulus $= 21$ kN/mm^2 and Poisson's ratio $= 0$. The loading per square metre of inclined surface area of each plate $= 3.2$ kN/m^2; this includes self-weight, allowance for finishes and snow load (wind can be ignored, see Section 9.2.3). *Table 9.6* gives information adequate for detailing schemes with eight, six and four plates respectively. In the table, Plate No. 2P8 means plate 2P of *Figure 9.16* and is for a

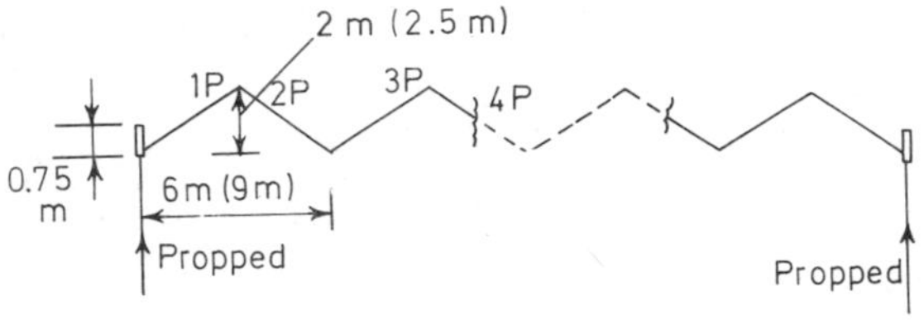

Fig. 9.16

scheme having eight plates. The notation is the same as in Section 9.1.5. The values of T_1 are for mid span and should be reduced towards the supports, using either a cosine or parabolic (like the bending moment for a beam) curve. The values of S are the maximum at the supports and are reduced linearly to zero at mid span. The values of M_2 are at mid span and can be taken as the same at all cross sections along the span. The values of T_2 are the maximum positive, or maximum negative if there are no positive values, anywhere in the plate. The values of R_V and R_H are the components of the reactions to the plates at the end transverse stiffeners and are used for designing these stiffeners. The values of M_1, N_1 and N_2 are the maximum irrespective of sign. These indicate that T_2, M_1, N_1 and N_2 produce very small stresses, which can either be ignored, or the design for T_1, S and M_2 made slightly conservative to allow for them. For T_1 and T_2 positive signs indicate tension. The forces are in kN/m run of plate and the bending moments are in kN m/m run of plate. For example for Plate No. 1P8, T_2 causes a tensile stress in the plate $= 11.3/0.1 = 113$ kN/m^2. The edge beams are propped by columns at say 5 m or less centres. For all the schemes the horizontal deflection of each edge beam, outwards, is 3 mm. For schemes with eight, six

Table 9.6.

Plate No.	Force or moment	Distance from left edge across width of plate, m								
		0.00	0.45	0.90	1.35	1.80	2.25	2.70	3.15	3.61
1P8	T_1	99.6	51.1	3.0	-45.1	-93.6	-143.1	-193.8	-246.4	-301.3
	S	-26.8	-32.1	-34.0	-32.5	-27.6	-19.3	-7.4	8.2	27.6
	M_2	-7.4	-4.3	-1.9	-0.2	0.8	1.1	0.8	-0.2	-1.9
	R_V	3.463	7.780	8.379	8.080	6.889	4.803	1.809	-2.111	-2.600
	R_H	5.507	12.06	12.7	12.09	10.23	7.079	2.615	-3.213	-3.904
	$T_2 = -11.3,$	$M_1 = 0.2,$	$N_1 = 1.2,$	$N_2 = 7.7$						
2P8	T_1	-301.3	-208.3	-117.2	-26.9	63.5	154.8	248.2	344.5	444.6
	S	27.6	45.6	57.1	62.2	60.9	53.2	39.0	18.0	-9.9
	M_2	-1.9	0.3	1.7	2.5	2.6	2.0	0.8	-1.1	-3.7
	R_V	4.190	11.18	14.17	15.53	15.27	13.34	9.724	4.346	-0.077
	R_H	-6.452	-16.96	-21.21	-23.06	-22.54	-19.63	-14.32	-6.541	-0.046
	$T_2 = 10.0,$	$M_1 = 0.3,$	$N_1 = 0.5,$	$N_2 = 6.6$						
3P8	T_1	444.6	323.9	206.8	92.2	-21.1	-134.2	-248.1	-364.2	-483.6
	S	-9.9	-37.1	-55.8	-66.4	-68.9	-63.4	-49.9	-28.3	1.7
	M_2	-3.7	-1.3	0.4	1.5	1.9	1.6	0.6	-1.0	-3.4
	R_V	2.411	9.120	13.92	16.59	17.19	15.75	12.28	6.781	1.042
	R_H	3.719	13.63	20.6	24.55	25.51	23.5	18.48	10.4	1.729
	$T_2 = 9.6,$	$M_1 = 0.4,$	$N_1 = 0.5,$	$N_2 = 6.2$						

4P8									
T_1	-483.6	-358.6	-236.9	-117.2	1.8	121.3	242.4	366.5	494.8
S	1.7	31.5	52.6	65.1	69.2	64.9	52.0	30.5	0.0
M_2	-3.4	-1.0	0.6	1.6	1.9	1.6	0.5	-1.2	-3.6
R_V	1.467	7.585	12.94	16.17	17.26	16.2	12.95	7.446	1.307
R_H	-2.378	-11.63	-19.47	-24.12	-25.6	-23.95	-19.14	-11.13	-2.081

$T_2 = 9.1, \quad M_1 = 0.4, \quad N_1 = 0.5, \quad N_2 = 6.2$

1P6									
T_1	99.4	51.0	3.1	-44.9	-93.4	-142.7	-193.4	-245.9	-300.7
S	-26.8	-32.1	-34.0	-32.5	-27.6	-19.3	-7.4	8.2	27.5
M_2	-7.4	-4.3	-1.9	-0.2	0.8	1.2	0.8	-0.2	-1.9
R_V	3.458	7.768	8.367	8.07	6.883	4.801	1.813	-2.099	-2.59
R_H	5.499	12.04	12.68	12.07	10.21	7.074	2.62	-3.196	-3.89

$T_2 = -11.3, \quad M_1 = 0.2, \quad N_1 = 1.2, \quad N_2 = 7.7$

2P6									
T_1	-300.7	-208.2	-117.5	-27.7	62.1	153.0	245.8	341.6	441.1
S	27.5	45.5	57.0	62.2	60.9	53.3	39.2	18.5	-9.2
M_2	-1.9	0.3	1.7	2.5	2.6	2.0	0.8	-1.1	-3.7
R_V	4.178	11.16	14.14	15.52	15.27	13.38	9.79	4.46	0
R_H	-6.433	-16.91	-21.17	-23.04	-22.54	-19.68	-14.42	-6.712	-0.158

$T_2 = 10.0, \quad M_1 = 0.3, \quad N_1 = 0.5, \quad N_2 = 6.6$

3P6									
T_1	441.1	322.3	207.1	94.4	-16.9	-128.0	-240.0	-354.0	-471.2
S	-9.2	-36.2	-54.9	-65.6	-68.3	-63.2	-50.2	-29.2	0.0
M_2	-3.7	-1.3	0.5	1.6	2.0	1.7	0.8	-0.8	-3.1
R_V	2.316	8.907	13.7	16.4	17.05	15.71	12.37	7.027	1.228
R_H	3.579	13.31	20.26	24.25	25.29	23.41	18.59	10.76	2.0

$T_2 = 9.6, \quad M_1 = 0.4, \quad N_1 = 0.5, \quad N_2 = 6.3$

Table 9.6. (continued)

1P4									
T_1	96.8	50.1	3.9	-42.4	-89.1	-136.7	-185.5	-236.1	-288.9
S	-26.0	-31.2	-33.1	-31.8	-27.1	-19.1	-7.7	7.2	25.8
M_2	-7.3	-4.2	-1.8	-0.1	0.9	1.3	1.0	0.0	-1.7
R_V	3.362	7.56	8.159	7.893	6.766	4.776	1.912	-1.845	-2.403
R_H	5.35	11.72	12.36	11.79	10.02	7.02	2.752	-2.828	-3.614

$T_2 = -11.1$, $\quad M_1 = 0.2$, $\quad N_1 = 1.2$, $\quad N_2 = 7.7$

2P4									
T_1	-288.9	-203.9	-120.8	-38.7	43.4	126.3	210.8	297.7	388.1
S	25.8	43.2	54.7	60.3	60.2	54.2	42.2	24.3	0.0
M_2	-1.7	0.4	1.8	2.6	2.6	2.0	0.8	-1.2	-3.8
R_V	3.938	10.61	13.58	15.07	15.09	13.59	10.56	5.931	1.011
R_H	-6.057	-16.06	-20.3	-22.36	-22.26	-20.01	-15.57	-8.9	-1.658

$T_2 = 10.0$, $\quad M_1 = 0.3$, $\quad N_1 = 0.6$, $\quad N_2 = 6.7$

and four plates the maximum deflections of internal valleys are 7, 4 and 3 mm respectively. The edge beams are theoretically propped continuously vertically; theoretically for the schemes with eight, six and four plates these beams have to withstand longitudinal tensions of 170.6 kN, 170.3 kN and 165.6 kN respectively, maximum sideways bending moments of 0.7 kN m in each case, vertical shear forces (due to change of bending moment giving complementary horizontal shear stresses) of 96.2 kN in each case and horizontal shear forces of 2.8 kN in each case. When R_V is positive this means that the transverse end stiffener is exerting an upward force on the plate. When R_H is positive this means that the reaction force provided by the end stiffener is from left to right.

Similarly to the above, *Table 9.7* gives information adequate for detailing schemes with four and two plates respectively, where $L =$ 25 m and the different values of height and width are shown in brackets in *Figure 9.16*. The edge beams are still 0.2 m wide and 0.75 m deep. They may need a 0.1 m square stiffener (like the one shown on *Figure 9.1*) at mid span to prevent the necessity for stress reduction (see CP 114, Table 14) in their top fibres. For the four- and two-plate schemes the maximum deflections of each edge beam outwards are 4.1 mm and 2.5 mm respectively. For the four-plate scheme the maximum deflection of the internal valley is 12.1 mm. The edge beams are theoretically propped continuously vertically; theoretically for the schemes with four and two plates these beams have to withstand longitudinal tensions of 274.5 kN and 197.2 kN respectively, maximum sideways bending moments of 0.7 kN m and 0.4 kN m respectively, vertical shear forces of 137.7 kN and 135.7 kN respectively, and horizontal shear forces of 3.5 kN and 2.5 kN respectively.

The methods of this chapter base design on elastic theory; the loads designed for are working loads and permissible working stresses (as in CP 114) should be used. This is because our main experience of shells and folded plates is based on elastic theory. With this experience large cracks do not develop at working loads and deflection calculations at working loads can be made, creep being allowed for by choice of Young's modulus for concrete (Section 2.3.15). It is difficult to do sufficient or even realistic research to justify ultimate collapse theories in a laboratory—this not only deals with ultimate strength of individual sections but plastic redistribution of forces and moments in three dimensions (for curved or doubly curved members in the case of shells). A cylindrical barrel vault shell say 18 m long and 63 mm thick scaled down for a laboratory test might be 1.8 m long, but then it would need to be 6.3 mm thick with perhaps four layers of reinforcement; shrinkage would not scale down, and would

Table 9.7.

Plate No.	Force or moment	Distance from left edge across width of plate, m								
		0.00	0.64	1.29	1.93	2.57	3.22	3.86	4.50	5.15
1P4	T_1	169.5	100.4	32.5	−35.0	−102.9	−172.0	−243.1	−317.1	−395.0
	S	−34.5	−45.4	−50.8	−50.7	−45.1	−34.0	−17.2	5.4	34.2
	M_2	−10.7	−5.0	−0.9	1.8	3.1	3.0	1.4	−1.6	−6.0
	R_V	5.809	13.84	15.78	15.88	14.17	10.65	5.3	−1.922	−3.851
	R_H	10.98	25.39	28.27	28.14	24.98	18.76	9.378	−3.281	−6.717
	$T_2 = -19.1,$	$M_1 = 0.3,$	$N_1 = 1.4,$	$N_2 = 10.1$						
2P4	T_1	−395.0	−280.5	−169.4	−59.9	49.3	159.9	273.3	390.9	514.3
	S	34.2	61.4	79.6	88.9	89.3	80.9	63.4	36.6	0.0
	M_2	−6.0	−1.2	2.1	4.0	4.5	3.6	1.2	−2.7	−8.1
	R_V	6.635	18.67	24.62	27.74	28.02	25.4	19.81	11.12	1.852
	R_H	−12.51	−34.3	−44.38	−49.47	−49.64	−44.86	−35.07	−20.14	−3.756
	$T_2 = 16.9,$	$M_1 = 0.4,$	$N_1 = 0.7,$	$N_2 = 9.6$						
1P2	T_1	123.3	82.0	41.8	2.2	−37.3	−77.2	−117.9	−160.0	−204.1
	S	−24.8	−33.1	−38.1	−39.9	−38.4	−33.8	−25.9	−14.7	0.0
	M_2	−9.7	−4.1	0.0	2.6	3.8	3.6	2.0	−1.1	−5.6
	R_V									
	R_H									
	$T_2 = -15.7,$	$M_1 = 0.2,$	$N_1 = 1.2,$	$N_2 = 9.9$						

be greater for the thin section, which could crumble under ancillary forces negligible to the large shell. The writer has supervised the work of Refs. 25, 26 and 27.

REFERENCES

1. Wilby, C. B., and Bellamy, N. W., *Elastic Analysis of Shells by Electronic Analogy*, Ed. Arnold (1962); Also published in Spanish by Compania Editorial Continental, S.A., Mexico, in metric units
2. Wilby, C. B., 'A proposed "exact" theory for analysing shells, and its solution with an analogue computer', *Proc. I. C. E.*, July (1962)
3. Wilby, C. B., and Khwaja, I., *Concrete Shell Roofs*, Applied Science Publishers Ltd, Amsterdam, London, New York (1977)
4. Part-time MSc course leaflet for structural engineering, Post-Graduate School of Studies in Civil and Structural Engineering, University of Bradford (1970)
5. Evans, R. H., and Wilby, C. B., *Concrete—Plain, Reinforced, Prestressed and Shell*, Ed. Arnold (1963)
6. Lundgren, H., *Cylindrical Shells*, Danish Technical Press, Inst. Danish Civil Engineers, Copenhagen (1951)
7. Tottenham, H., 'A simplified method of design for cylindrical shell roofs', *The Structural Engineer*, June (1954)
8. Wilby, C. B., 'A Method of Designing North-Light Shell Roofs', *Indian Concrete Journal*, Jan. (1961)
9. Schorer, H., 'Line load action on thin cylindrical shells', *Proc. Amer. Soc. Civil Engrs.* (1935)
10. Jenkins, R. S., *Theory and Design of Cylindrical Shell Structures*, O.N. Arup Group of Consulting Engineers (1947)
11. Wilby, C. B., *Elastic Stability of Post-tensioned Prestressed Concrete Members*, Ed. Arnold (1964)
12. Rao, V. M., *Elastic Analysis of Continuous Cylindrical Shell Roofs*, PhD Thesis, University of Bradford (1976)
13. Wilby, C. B., and Naqvi, M. M., *Reinforced Concrete Conoidal Shell Roofs— Flexural Theory Design Tables*, Cement and Concrete Association (1973)
14. Goldberg, J. E., and Leve, H. L., *Theory of Prismatic Plate Structures*, International Assocn. Bridge and Structural Engineering (1957)
15. Goldberg, J. E., Glauz, W. D., and Setlur, A. V., *Computer Analysis of Folded Plate Structures*, International Asscn. Bridge and Structural Engineering, Rio de Janiero, 16 Aug. (1964)
16. Gibson, J. E., and Gardner, N. J., 'Investigation of Multi-folded Plate Structures', *Proc. I.C.E.*, May (1965)
17. Winter, G., and Pei, M., 'Hipped Plate Construction', *Journ. Amer. Conc. Inst.*, Jan. (1947)
18. Gaafar, I., 'Hipped Plate Analysis Considering Joint Displacements', *Proc. Amer. Soc. Civil Engrs.*, Apr. (1953)
19. Rockey, K. C., and Evans, H. R., 'A Critical Review of the Methods of Analysis for Folded Plate Structures', *Proc. I.C.E.*, June (1971)
20. Simpson, H., 'Design of Folded Plate Roofs', *Journ. Struct. Div. Amer. Soc. Civil Engrs.*, Jan. (1958)
21. Parme, A. L. L., 'Direct Solution of Folded Plate Concrete Roofs', *Bull. I.A.S.S.*, No. 6 (1960)

22. Thadani, B. N., 'The Analysis of Hipped Plate Structures by Influence Coefficients', *Indian Concrete Journal*, Apr. (1957)
23. Tamhankar, M. G., and Jain, R. D., 'Computer Analysis of Folded Plates', *Indian Concrete Journal*, Oct. (1965)
24. Parme, A. L. L., 'Computational Arrangement for Analysis of Folded Plate by Direct Solution', *Advanced Engineering Bulletin*, 3a, Portland Cement Assoc., U.S.A. (1963)
25. Khwaja, I., *Theoretical Analysis and Experimental Behaviour of Hyperbolic Paraboloidal Shells*, PhD Thesis, University of Bradford (1968)
26. Naqvi, M. M., *Theoretical Analysis and Experimental Behaviour of Conoidal Shells*, PhD Thesis, University of Bradford (1969)
27. Husain, I., *Structural Analysis of Conoidal Shells and Frames using Direct Models*, PhD Thesis, University of Bradford (1974)

Index

Adhesion, 40
Age of concrete, 15, 19, 29
Aggregate:cement ratio, 23
Aggregates, 9–12
  fine, coarse, 9
  gradings, 10, 23–28
  lightweight, shape, 10
  rounded, irregular, angular, 11
Air entrainers, 14, 141
Anchorage, 40–52
  splitting effects of, 55, 56
Anchorage lengths, 42–52, 135
  elastic theory, 56
Anchorage of column bars into bases,
  135, 136
Anchorage of shear bars, 52
Anchorage of stirrups, 54, 55
Arches, 165, 166
Asbestos cement, 4

Balanced design, 87, 90, 93
Barrel valult roofs, 166 (see *Cylindrical
  shell roofs*)
Basement wall, 5
Bases, 132–135
Batching concrete by volume and weight,
  11, 12
Beam
  depth, 58, 60, 98
  breadth, 58
  T- and L-, 58
Bearing stresses at bends, 53
Bending moments, redistribution, 121,
  222
Bogue compounds, 3
Bond, 40–52, 60
Bond stress, 30, 42, 82
Briquettes for tension test, 15

Calcium chloride, 7, 8, 19, 144

Cement,3
  extra rapid hardening, 5
  for cold weather, 7
  testing of, 15
  with low coefficient of shrinkage, 8
  with low heat of setting, 8
Characteristic load, 2
Characteristic strength, 2, 22
Characteristic stress, 2
Chemical conversion, 6, 7
Ciment Fondu, 4, 5
Coloured cements, 9
Columns, 111–120
  axially loaded short, 111
  circular, 119
  eccentrically loaded, 112–118
  short, 111
  slender, 111
Compacting factor, 13
Composite construction, 168
Compression steel, 94–97
Compression steel near neutral axis, 96
Concrete
  quantities of materials, 28
  use of, 13
  voids in, qualities, compaction, 12, 23
Continuous beams, 121, 122
Corrosion of tendons, 143, 144
Cover of concrete, 66
Cracks, 31, 60, 98, 100, 108, 109, 141
Creation of structures, 38–40
Creep, 18, 32–35, 145, 157
Curing of concrete, 19
Curtailment of reinforcement, 50, 51
Cylinder splitting test, 15

Deflection, 1, 59, 97, 100, 108
Design of bases, 132–135
Design of beams, 58, 126–130
Design of columns, 112, 118–120, 130–
  132

Design of compression steel, 95
Design of conoidal shell roofs, 188–191
Design of continuous cylindrical shells, 186–188
Design of floor of building, 124–130, 136
Design of folded plate roofs, 191–207
Design of frames, 121, 122
Design of hollow tile floor, 128
Design of prestressed concrete members, 147–170
Design of shear reinforcement, 76
Design of slabs, 100, 101, 124–126, 136
Design office student exercise, 136
Design strength, 2
Designers' tables
  anchorage lengths, 43, 44, 47, 56
  areas of bars, 69
  areas of bars for slabs, 69
  balanced design values K_1 and ρ, 91
  bending moments and shear forces in continuous beams and slabs, 122, 138
  bending moments, support reactions and deflections for beams with fixed and free supports, 137
  bent-up bars, 81
  curtailment of bars, 48
  elastic analysis of conoidal shells, 188–191
  elastic analysis of cylindrical shells, 186–188
  folded plates, 201–207
  folded plates (Parme), 193–200
  hooks and nibs, 46, 47
  ratios of span to overall depth, beams and slabs, 126
  shell reinforcement quantities, 178
  stirrups, 80
  stress in compression steel, 95
  values of z_1 and $K = M/(bd^2)$ for elastic theory, 70
  weights of materials, 139, 140
Diagonal tensile stresses, 74, 76
Dilatency, 41
Ducts, tubing and inflatable, 142
Durability, 14

Economics, 38–40
Elastic theory/analysis, 1, 2, 56, 59, 71
Elastic theory formulae for slabs and beams, 67
End anchorages, 44

Epoxy resin, 142
Equivalent area, 62

Factor of safety, 1, 59
Factor γ_m, 2
Flash set, 4
Flat slabs, 100, 101, 136
  dropped panels, 101
Flexural compression failure, 87
Fly ash, 8
Folded plate roofs, 166, 173, 191–207
  analysis due to Parme, 193
  design, 191, 192
  design tables, 194–200, 201–207
Fourier's series, 184
Frames, 101, 121, 122
Friction, 41, 146, 147, 157
Frost resistance, 14, 141

Gap graded concrete, 17, 18
Grading, combing aggregates, 26–28
Granolithic, 31
Gravity, acceleration due to, 140
Grip, 40
Grouting ducts, 141

Hardening, rate of, 4
High alumina cement, 4, 6, 19
Hollow-tile floor, 128
Hooks, 44–46
Hyperbolic paraboloids, 166

Laps in reinforcement, 47, 49
Limit state design, 1
Load factor, 1
Loading tests, 108
Losses of prestress, 144–147, 156, 157

Mean strength, 22, 23
Mix design, 17, 20–28
Modular ratio, 61, 62
Modulus of elasticity, 33–35, 61
Modulus of rupture, 60
Moment of inertia, 61, 62, 64
Moment of resistance, 64, 66

Neutral axis, 61, 64

Nibs, 44–46
No fines concrete, 18, 19

Over-reinforced, 87

Partial prestressing, 148
Permissible stresses, 1
Plastic analysis, 86
Plastic analysis assumptions, 86, 87
Plastic design in bending, 87
Plasticisers, 13, 141
Poisson's ratio, 60
Polyester, 142
Portland blast furnace cement, 9
Portland cement, 3
  ordinary, rapid hardening, 5
Portland-pozzolana cement, 8
Post-tensioning, 141
Post-tensioning for shear, 142
Prescribed mixes, 28
Prestressed concrete, 6, 8
  additional untensioned steel, 163
  advantages and disadvantages, 142, 143
  assumptions for elastic design, 151, 152
  classes of structures, 148
  columns, 165
  compression steel, 164
  continuity, 169
  end splitting forces, 169
  inclined tendons, 167, 168
  limit states of deflections and stresses, 152–155
  losses, 31, 144–147, 156, 157
  materials, 143
  shear resistance, 166–168
  tanks, pipes, domes, shells and piles, 169, 170
  ties, 165, 166
  torsional resistance, 170
  ultimate limit state due to flexure, 158–165
Prestressing beds, 166
Prestressing dams, 142
Pretensioning, 141
Principal stresses, 166
Pull-out test, 41

Reinforcement, 35–38
Relaxation, 38, 144, 156

Serviceability limit state, 1
Set: initial, final, 4
Setting time: initial, final, 4
Shear reinforcement, 76–82
Shear strength, 16
Shear stresses (elastic), 71, 75
Shell roofs, 172
  conoidal, 188–191
  continuous, 186–188
  cylindrical (symmetrical and North-Light), 172–188
  design, 172, 173, 179
  economics of, 172, 173
  elastic analysis, 179–186
  electronic analogue, 185
  estimating reinforcement in, 178, 179
  proportioning of, 173–178
  Schorer, 182, 185
Shrinkage, 5, 29–31, 145, 156, 157
Shrinkage stresses, 60, 74, 76
Slabs
  Hillerborg's strip method, 100, 106–110
  holes in, 104, 108
  isotropically reinforced, 101, 102
  Johansen's yield line, 100–106
  on solid, 31
  orthogonally anisotropically reinforced, 103
  orthotropically reinforced, 103
  skew, 108
  spanning one way, 100
  spanning two ways, 100
Slip, 40, 145, 157
Slump test, 12, 13
Space frames, 165
Steam curing, 6, 20, 147
Stopping-off bars, order of, 48
Strain loss, 144
Strand, 38
Strength of concrete, 5, 14, 19
Strength tests, 15
Stress-block, 87
  CP 110, 89, 90, 94, 95, 160–165
  Whitney, 88, 89, 94, 95, 158, 160
Stress: strain relationship, 32, 159–161
Strip method, Hillerborg's (see *Slabs*)
  CP 110, 109
  discontinuity lines, 108
Sulphate resisting cement, 8

T-beams, 96

Tables for designers, see *Designers' tables*
Tank for water, 68
Tanking, 5
Temperature stresses, 60, 76
Tobermorite gel, 5
Torsion, 83–86
Torsion test, 16
Truss-analogy, 76

Ultimate strength, 1, 2
Under-reinforced, 87, 88, 94
Units, conversion British Imperial and U.S.A. units to SI, ix

Vacuum concrete, 16
VB consistometer test, 13
Vibrated concrete, 16

Walls, 118
Water-retaining structures, 2, 56, 59, 60, 66, 68, 71
Water-to-cement ratio, 5, 14, 15, 24, 25, 27
Wedge action, 41
Whitney, see *Stress block*
Workability, 12, 13, 18, 23–25, 27, 142
Working loads, 1, 59
Working stress, 1

Yield line, Johansen's, see *Slabs*
    corner levers, 106
    CP 110, 109
    design of bases, 133, 134
    equilibrium method of analysis, 101
    kinking of reinforcement, 105
    membrane action, 105
    upper bound solutions, 105
    virtual work method of analysis, 102